AMERICAN INSTITUTE FOR ECONOMIC RESEARCH

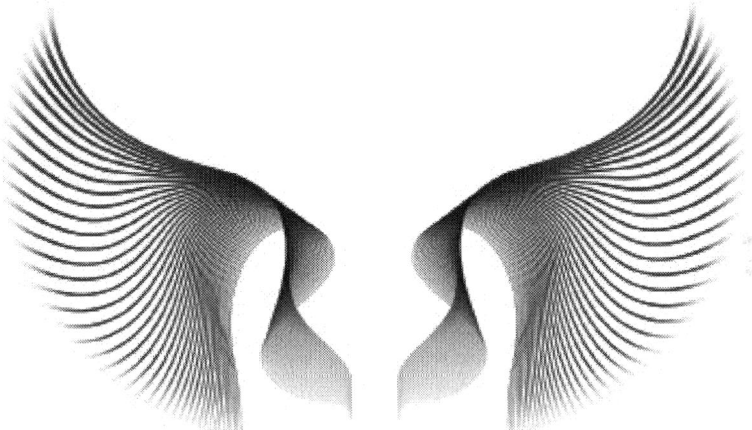

THE NEW TECHNOLOGIES OF FREEDOM

DARCY W.E. ALLEN · CHRIS BERG · SINCLAIR DAVIDSON

The New Technologies of Freedom

By Darcy W.E. Allen, Chris Berg and Sinclair Davidson

ISBN: 9781630692070

Cover art: Vanessa Mendozzi

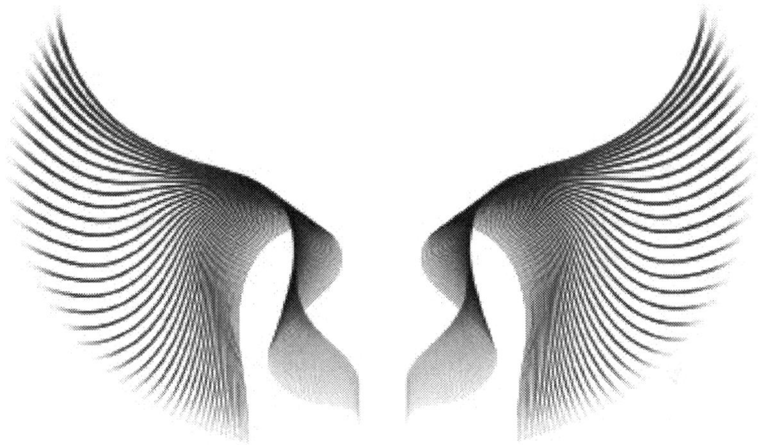

THE NEW TECHNOLOGIES OF FREEDOM

DARCY W.E. ALLEN - CHRIS BERG - SINCLAIR DAVIDSON

AIER | AMERICAN INSTITUTE for ECONOMIC RESEARCH

CONTENTS

PREFACE AND ACKNOWLEDGMENTS

This book comes at a time of deep technological and political change. Advances in blockchains, artificial intelligence, privacy technologies and many others, have elevated public fears about how technology affects our freedoms. We wrote this book before the pandemic of 2020, but that crisis has only accelerated rapid technological adoption - by both citizens and the state. Long before the pandemic we were in the middle of a global debate about freedom of speech, privacy and autonomy in a digital world. Those issues, together with each of our interests in the pursuit of liberty, have motivated this book. Techno-pessimism is not only misguided, but harmful. The central theme of this book is the belief that technologies are tools to expand individual freedoms, prosperity and satisfaction—and the liberty movement should adjust its strategy accordingly.

In his 1983 book, *Technologies of Freedom*, Ithiel de Sola Pool outlined how the new communications technologies that were then emerging in the developed world had radical implications for the traditional government controls over speech and expression.[1] The advent of cable television and electronic publishing revolutionised the environment for freedom of speech, allowing individuals to circumvent and undercut censorship and regulation. Pool's book was published at only the very start of the revolution he described. The waves of communication innovation since have completely broken down the barrier between content makers and their audience. They have not only increased our ability to speak to each other, but they have materially increased our freedom to speak.

We write this book, *The New Technologies of Freedom*, with a similar goal: that technologies we describe here are only the start of a broader revolution in liberty—that they offer an historically unparalleled opportunity to take control of individual freedom.

This book was jointly conceived and co-designed with Ron Manners, and the Mannkal Economic Education Foundation of which he is founder and chairman has generously supported the project. Ron is a significant figure within the liberty movement not just in Australia but around the world. Western Australia is one of the great hubs of liberty entirely due to the tireless effort of Ron and his team in ensuring that young people have both the opportunity and the resources to experience and contribute to the liberty movement. There is barely a liberty-dedicated organisation on the planet that has not hosted one of his young Mannkal scholars.

This project has come about because we share Ron's optimism about the future of liberty and especially his desire to take the great ideas in the liberty tradition and make them real. In a speech before

---

1 de Sola Pool, Ithiel. 2009. *Technologies of Freedom*. Cambridge, MA: Harvard University Press.

the Mont Pelerin Society meeting in Gran Canaria, Spain, in 2018, Ron reflected on spending time with one of our shared heroes, the economist Friedrich Hayek:

> Prof Hayek reminded me of the important role that entrepreneurs should be playing in the battle of ideas. Once, I asked him to "slow down because I'm not an economist," to which he replied, "I'm glad you are not an economist. We economists simply dream our ideas and think our thoughts but you go out and — fire the bullets!" I walked away from that meeting feeling that I had been given a useful role in society …
>
> We now have the ammunition so the focus must be on strategic execution, rather than simply maintaining our 'Remnant' role. We may have fed the troops from the comfort of our ivory towers but it is now time to move to the front line.
>
> We need to experience the exhilaration of firing the intellectual bullets and decisively winning by focusing on what we can offer to individuals, rather than nations, by developing 'bottom up' strategies.[2]

This book, and the project which it forms a part of, is a response to that need—to show to those individuals who wish to take back their liberty how they may do so. We seek to redirect the focus of our movement away from complaining about governments, and to governments, and towards empowering individuals. Ron's enthusiasm for the future of the liberty movement is infectious, and we hope that this

---

2 Manners, Ron. "Liberty could be good for you too!" *Speech at Mont Pelerin Society*, Gran Canaria, Spain, October 23, 2018.

book contributes to the spread not just of the ideas of freedom, but of the experience of freedom. We would also like to strongly thank Ron's team—particularly Andrew Pickford for his longstanding support, and Kate Wagstaff for her valuable help in the early stages of the project.

This book is a product of an on-going research program on the economics, law and policy of frontier technologies at the RMIT Blockchain Innovation Hub. The RMIT Blockchain Innovation Hub was born out of our shared interest for a wide range of new technologies—including blockchains, 3D printing and privacy technologies—and our use of economics to understand their impact on our lives. It is strongly supported by RMIT University and the College of Business and Law. For this reason, many of the ideas in this book have come from insightful discussions with colleagues and associates of the RMIT Blockchain Innovation Hub, including Jason Potts, Aaron Lane, Vijay Mohan, Marta Poblet, Alastair Berg, Stuart Thomas, and Lachlan Tierney. We have also canvassed the ideas with other teams including Agoric (including Mark Miller, Dean Tribble, Bill Tulloh and Kate Sills), the International Centre for Law and Economics (Geoffrey Manne, Julian Morris, and Gus Hurwitz), and the Worldwide Blockchain Innovation Association (Stuart Eaton).

This book is also based on our active involvement in the liberty movement as academics, public commentators, and in the public sphere. Over many years we have written books, academic papers and opinion pieces, and attended conferences on technology and liberty. We have made hundreds of media appearances expanding and advocating the ideas of freedom, technology and entrepreneurship. In a very real sense our engagement with the liberty movement—its thinkers and activists—has brought us to these conclusions. While a full accounting of the remarkable individuals and teams that have informed, challenged and inspired our thinking on this book is not possible, we would like to thank our current and former colleagues at the Institute of Public Affairs (led by John Roskam), the American Institute for Economic Research (Jeffrey Tucker and Ed Stringham), and the Competitive Enterprise Institute (Iain Murray for commissioning some of our thinking on the future of privacy).

There are many other people and organisations we would like to thank, including Mikayla Novak, Trent MacDonald, Brendan Markey-Towler, James Caton, Eric Alston, Pavel Kuchar, Keith Hankins, Joe Clark and Peter Rohde. We have benefited greatly from the editorial teams and anonymous reviewers at the Journal of Special Jurisdictions and the Review of Austrian Economics, as well as editors at Palgrave MacMillan and Edward Elgar. We have also presented these ideas widely including at the Adam Smith Institute, Australian Libertarian Society Conferences in 2019 and 2020, Mont Pelerin Society 2017, RMIT Blockchain Proof of Work Seminar Series, Linked Democracy: AI for Democratic Innovation Conference and the International Society for the Study of Decentralized Governance meetings.

With this book we hope not only to shift the public towards technology-optimism, but also to shift the strategy that the liberty movement takes in the digital economy of the twenty-first century. We argue that the liberty movement needs to spend less time begging the state for its freedoms back. Rather, those interested in expanding liberties need to be more entrepreneurial: building the monies, institutions and organisations that make us freer. As we outline in this book, technologies can be tools for economic, political and social freedom.

01.

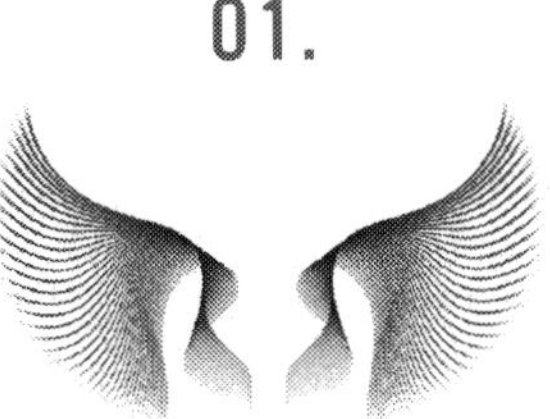

INTRODUCTION

This book is about technology, individual freedom and optimism—about how a suite of new technologies is poised to enhance human liberty and flourishing. Our claim is that technological developments in blockchain, automation, artificial intelligence and machine learning, advances in cryptography and the digitisation of ownership expands individual rights, the protection and expansion of property rights, the defence of privacy and personal data, and even human satisfaction.

Our hope is that the book provides a guide for the recapturing of liberties in the digital age. There is a pressing need to empower individual citizens with the ability to take control of their own rights and freedoms, even in the absence of government support. This need is especially deep in environments where governments and policymakers are hostile to individual liberty.

One or two decades ago this optimistic stance about the relationship between technology and liberty would not have been unique. Readers in the 1980s, 1990s and even 2000s had available at their hands a stream of books, articles, and pronouncements about the liberating power of the communications revolution. This was the age of *Wired* magazine, which offered readers in the vibrant Silicon Valley and around the world a taste of not just techno-optimism but sometimes a sense of techno-utopianism. Its covers and essays proclaimed everything from the end of advertising to the "primordial stirrings of a new kind of nation—the Digital Nation—and the formation of a new post-political philosophy."[1]

This was the age when the just retired Ronald Reagan could declare, watching the demise of the Soviet Union that "the biggest of Big Brothers is increasingly helpless against communications technology," that "electronic beams blow through the Iron Curtain as if it were lace," and that the "Goliath of totalitarianism will be brought down by the David of the microchip."[2] In his 1992 book *The End of History*, Francis Fukuyama described the creation of a universal consumer culture driven by communications technology—in his case, the VCR.[3]

Fukuyama was right: when the soap opera *Dallas* was aired in communist Romania as an exemplar of the ethical wrongs of capitalism in the West, viewers instead marvelled at the luxuries and prosperity

---

1 Karpf, David. 2018. "25 Years of Wired Predictions: Why the Future Never Arrives." *Wired*, September 18.

2 Rule, Sheila. 1989. "Reagan Gets a Red Carpet from the British." *The New York Times*, June 14.

3 Fukuyama, Francis. 2006. *The End of History and the Last Man.* New York, NY: Free Press.

apparently available to their ideological opponents.[4] Estonian viewers, tuning into Finnish television across the border marvelled at the wealth portrayed in shows like *Dallas*, *Knight Rider*, and the Finnish advertisements that wrapped around them.[5] Technology was a force for liberation.

Few are as unabashedly optimistic now. Initially, the Arab Spring of 2010 (and the Iranian protests that preceded it) seemed to be a fulfilment of the promises of techno-optimists of the past decades; this was the revolution that would be tweeted. Hillary Clinton, then Secretary of State in the Obama administration, gave a stirring speech at the Newseum on how the internet is both "helping people discover new facts and making governments more accountable."[6] But the Arab Spring gave way to conflict across the region and a resurgence of authoritarianism. Tunisia and Egypt went for stability rather than liberty, Syria and Libya descended into civil war, and in the power vacuums created by American interventionism in Iraq grew a new generation of Islamic radicals able to wreak havoc across the region.

The dream of a revolution by social media was far away. Authoritarians outside the Middle East, observing this uproar, responded with crackdowns of their own. Just two years after the Olympic games that

---

4 Gillespie, Nick, and Matt Welch. 2008. "How 'Dallas' Won the Cold War." *The Washington Post,* April 27.

5 Lepp, Annika, and Mervi Pantti. 2013. "Window to the West: Memories of Watching Finnish Television in Estonia During the Soviet Period." *View Journal of European Television History and Culture* 1 (3): 77-87; Holden, Stephen. 2010. "J. R. Ewing Shot Down Communism in Estonia." T*he New York Times*, November 11.

6 Clinton, Hillary. 2010. "Remarks on Internet Freedom." *Exhibit in The Newseum*, January 21. The Newseum, for its part, a museum dedicated to journalism, founded by a non-profit focused on the First Amendment, is scheduled to close in January 2020 in the wake of financial difficulties.

were heralded as its own version of *glasnost* the Chinese reasserted the power of its state—there would be no Jasmine Revolution in a country whose Great Firewall was configured to block even the words "Egypt" and "Tunisia."[7] China has pivoted towards high-tech authoritarianism in the decade since. Its investments in artificial intelligence and the internet of things have built a surveillance and control state without historical parallel. Not only have the technologies of liberty been revealed as next to useless in the face of concerted suppression, the digitisation of everyday life in China has allowed for greater, and more overbearing, government.

In the developed world too, these new digital technologies have been revealed to have a dark side. Social media was first treated as a curiosity (why would anyone post pictures of their breakfast?) then as a welcome sign of democratic disintermediation (we could interact directly with famous artists, journalists and politicians), but now it is increasingly seen as fuelling polarisation, fostering disinformation and propaganda, and undermining personal privacy. The sudden prominence of Facebook (and to a lesser extent, Twitter and Facebook's subsidiary photo-sharing site: Instagram) in the lives of citizens in almost every demographic has placed it at the centre of politics, culture, and the media. At the same time these platforms have emerged as the facilitator of our most personal relationships. Facebook alone offers a universal platform on which we can engage our family, our friends, our colleagues, our democratic representatives—and all this for a product that only became widely available in 2008.

The explosion of "internet of things," or IoT, devices means that

---

7 Richburg, Keith B. 2011. "Nervous About Unrest, Chinese Authorities Block Web Site, Search Terms." *The Washington Post*, February 25.

everywhere we travel there are sensors monitoring the environment around us. Our phones record where we have been and the stores we visit. Our cars, our computers, our fridges, our smart speakers and lights and air conditioners stream data back to their manufacturers. Facial and license plate recognition software tracks us as we travel through the urban environment. The loss of privacy—that sensation of being unobserved, being anonymous in the crowd—is palpable.

Now, as we write this book, the social-technological-political landscape is different. There are drum beats for the heavy regulation of new technology being offered by government bodies and political candidates alike. The Australian Competition and Consumer Commission called for a regulatory agency that would monitor the algorithms of social media platforms. Presidential candidates have called for the breakup of Facebook. Conservatives frustrated by "deplatforming" have described access to these networks as a "civil rights issue" akin to Jim Crow-era racial discrimination.[8] Ridesharing, food delivery and digital labour hire services have come to be targeted by unions and other critics of the gig economy, who call for more regulation and decry the loss of benevolent management. The home-sharing service AirBnB is blamed for making some of the world's most prominent cities unlivable—Barcelona, Berlin, New Orleans, San Francisco, London——as it boosts tourist demand and comes up against anti-growth planning laws. The social cache of technology—the gloss that used to accompany the Silicon Valley brand—has fallen away. We drank in techno-optimism and the hangover is deep.

Of course, this optimism was never as dominant as its critics have

8 Chamberlain, Will. 2019. "Platform Access Is a Civil Right." *Human Events*, May 3.

claimed. In its very first issue, *Wired* described the "Digital Revolution … whipping through our lives like a Bengali typhoon."[9] This was hardly an unambiguous statement of the positive effects of technology. Google's famous 'don't be evil' slogan had embedded within it a critique of Google's competitors. Understood in the context the motto was coined, to 'be evil' would be to bias search results in favour of advertisers, implying that other firms had.[10]

The tech revolution has always needed—and outside the clichés about Silicon Valley or the "California ideology," has always known that it needed—business ethics and philosophy.[11] So too does the finance sector, the supply chain and transportation industry, food and agriculture, manufacturing, services and hospitality. But each of those industries has evolved over the space of many decades, if not hundreds of years. The digital sector has gushed forth in less than two. The industrial revolution took at least three times that, and multiples again before it touched every corner of society.

The backlash, then, against the technology industry is explicable—a widespread realisation that disruption is disruptive, that revolutions are unruly, that progress is neither linear nor certain. But it is just as much of an error to reverse the stance; to replace optimism with pessimism, utopianism with dystopianism. The argument of this book is that techno-optimism has never been more relevant.

But our aim here is not simply to survey the future of technology

---

9 Rossetto, Louis. 1993. "The Wired Manifesto." *Wired*.

10 Page, Larry, and Sergey Brin. 2004. "Letter from the Founders: 'An Owner's Manual' for Google Shareholders." August 18.

11 Barbrook, Richard, and Andy Cameron. 1995. "The Californian Ideology." *Mute*, September 1.

and freedom. We have two additional aims in mind. First, to restore optimism to the technology debate. We're excited about the future and we think you should be as well. Second, to urge you to develop that future—to experiment with these new tools and opportunities, to use them to take back your freedoms, to build and to share what you have found with the rest of us. At the end of the 2010s so many of us have so thoroughly imbibed the idea that the technological future is not made by us, but is something that happens to us. But building the future is a communal task.

What do we mean by the new technologies of freedom? These are a set of technological innovations in various stages of production and development that we argue will, if adopted widely, result in more responsive markets, enhanced self-autonomy, better protection for our freedoms, and more opportunity and prosperity.

We believe this optimism ought to be shared by readers of all political stripes. In our work at the RMIT Blockchain Innovation Hub, a research centre dedicated to the social science of blockchain technology, we have found ourselves working with self-professed libertarians, conservatives, social democrats, and Marxists in developing out new models of charities, of cultural production (music, film, art), of trade and supply chains, of aid and development, of distributed governance and distributed economic production. The philosophical diversity of the blockchain community is much greater than usually understood. But that community is bound together by a shared belief that this is technology a force for liberation, for the dismantling of hierarchies, whether private or public, and for deeper and better economies and societies.

A final thought about what this book is not. It is not a defence of the technology industry, or Silicon Valley, or an argument that things are not as bad as they seem. The case for pessimism that we outlined above is not always compelling. For instance, the impact that AirBnb is having on cities is almost entirely because of those cities' planning restrictions, that prevent entrepreneurs bringing more supply into the market for accommodation to meet demand. Heavily regulated sectors find it hard to adjust to change—but the problem there is the regulation, not the change. Yet on the other hand there are real problems about data privacy, the internet of things and cyber-security, and the authoritarian use of artificial intelligence by governments.

One of us has described the "social negotiation" between firms, consumers, the technology industry and, indeed, government, that characterises the early stages of all new technologies.[12] Just as markets do not instantly recalibrate after shocks, new rights and philosophies must be discovered. This book is part of the negotiation. But this book is also a call for a shift in focus for those who care about individual freedom. These new technologies present a historically unparalleled opportunity for entrepreneurs to build institutions of freedom, rather than just arguing or voting for legislative change.

The liberty movement—the global network of academics, activists, policymakers, philanthropists, and engaged citizens with the shared ideals of free markets and individual rights—spends too much of its resources begging the state for its freedoms back. The usual strategic thinking is this: liberties are limited through legislation. So it is legislation that has to change. Bills that harm individual freedom must be rejected, and bills that increase freedom promoted. The main business of a typical liberty-focused think tank is producing research into

---

12 Berg, Chris. 2018. *The Classical Liberal Case for Privacy in a World of Surveillance and Technological Change*. Palgrave Macmillan.

specific public policy issues, proposing reforms that would enhance freedom and, and opposing new bills that would reduce freedom. Liberty-minded scholars measure how effective or ineffective legislation has been and identify new directions for deregulation. Activists send mass emails to their representatives to support or condemn legislative action.

This focus on legislative reform is baked into the self-image of much of the liberty movement. The Heritage Foundation is the largest conservative think tank in Washington DC and one which has a particular emphasis on activism and political action. Heritage places a dispute about exactly how to best effect legislative change at the heart of its founding story.[13] In 1971, two Republican staffers were surprised to receive copies of a publication by the American Enterprise Institute on US government funding for a supersonic transport plane two days after a crucial vote on the topic. Frustrated that they had not received it sooner, they inquired as to why. The AEI president allegedly said "We didn't want to try to affect the outcome of the vote." The two staffers—Ed Fuelner and Paul Weyrich—founded the Heritage Foundation two years later to take a more direct approach. Success would now be measured on direct policy change.

Other strategic approaches of course exist. Liberty movement activists increasingly focus on direct public engagement, not only through the traditional mediums of television and print, but social media, accessible videos and podcasts that target engaged popular audiences. Milton Friedman is the archetypal free market public intellectual, and his 1980 book *Free to Choose*, and its accompanying television series, are the archetypal instance of the popularisation of

---

13 See: Edwards, Lee. 2013. *Leading the Way: The Story of Ed Feulner and the Heritage Foundation.* Crown Publishing Group.

free market economics. The Federal Reserve governor Ben Bernanke has emphasised how Friedman "conveyed to millions" his beliefs in open markets and the liberal order.[14] The liberty movement now has its own awards for public commentary and journalism (the *Reason* magazine media awards), for popular video and commentary (the Atlas Network's Lights, Camera, Liberty award), and strategic impact (the Templeton Award, also by Atlas Network). That latter award specifically identifies the need for award winners to have "laid the groundwork for future progress in improving countries' scores in rankings of economic freedom."[15] But these public engagement approaches also have the implicit goal of ultimately shifting the terms on which democratic politics—that is, the politics of legislative change—are fought.

It is absolutely the case that the liberty movement has had a great deal of success by focusing on public opinion and legislative action. The liberty movement's opponents certainly believe so. There are a host of books and articles arguing that the free market reforms—deregulation, privatisation, tariff liberalisations and so forth—since the 1970s were inspired or led by the think tank, advocacy and political networks of the libertarian and conservative centre-right. Naomi Klein, for instance, described an "ideological crusade" led by Milton Friedman and Washington DC think tanks to usher in privatisation in the wake of military action and crisis.[16] A more serious volume edited by Philip Mirowski and Dieter Plehwe described the path from the academics of

---

14 Bernanke, Ben. 2002. "On Milton Friedman's Ninetieth Birthday." *Conference to Honor Milton Friedman, University of Chicago, Chicago, Illinois.*

15 Atlas Network. 2019. "The Templeton Freedom Award."

16 Bernanke, Naomi. 2007. *The Shock Doctrine: The Rise of Disaster Capitalism*. Toronto, Canada: Alfred A. Knopf.

the Mont Pelerin Society—an intellectual society founded by, amongst others, Friedrich Hayek—to the market reform throughout Europe and the Americas.[17]

National histories tell the same global story. Ronald Reagan declared that the historians of the twenty-first century would focus on the role played by the Heritage Foundation—the think tank that launched a conservative intellectual revolution—in shaping elite political thought about economic reform.[18] In Australia, the fact that the economic reform movement was started by Labor governments—either dating to the 1973 tariff reductions of the Whitlam government, or the floating of the Australian dollar by the Hawke government a decade later—has not prevented the free market movement being credited for changing the intellectual zeitgeist. The journalist Paul Kelly, in his magnum opus *The End of Certainty* about the politics of the 1980s, focused on the influence of the two free market think tanks the Institute of Public Affairs and the Centre for Independent Studies, in driving this reform.[19] A more recent book has focused on the lobbying and research efforts of smaller, issue-focused organisations that called for industrial relations changes, and protection of property rights in the wake of indigenous land title, and opposed climate mitigation policies.[20]

---

17 Mirowski, Philip, and Dieter Plehwe. 2015. *The Road from Mont Pèlerin*. Cambridge, MA: Harvard University Press.

18 Stedman Jones, Daniel. 2012. *Masters of the Universe: Hayek, Friedman, and the Birth of Neoliberal Politics*. Princeton, New Jersey: Princeton University Press.

19 Kelly, Paul. 1992. *The End of Certainty: The Story of the 1980s*. *St Leonards*, NSW: Allen & Unwin.

20 Kelly, Dominic. 2019. *Political Troglodytes and Economic Lunatics: The Hard Right in Australia*. Melbourne, Australia: Schwartz Publishing.

We are the better part of half a century since the reform movements of the 1970s and 1980s. The free market movement can take credit for some major successes in that time. But while the public policy environment has markedly changed, the strategic approach that focuses on public opinion and legislative change has come up against what seems to be some firm limits.

Take, for instance, the size of government. Across the developed world, levels of government spending and taxation has been relatively steady over the past half century. For the United States, federal government spending has hovered consistently between 17 percent of GDP and 22 percent of GDP, with only a sharp spike occurring in the wake of the Global Financial Crisis. The United Kingdom (which has a unitary rather than federal system of government) has been more volatile over the same period, yet no more optimistic. While there was a sharp decline in the reform period in spending (from a spending peak of 46 percent of GDP in the late 1970s, spending was pulled down to as low as 34 per cent of GDP in the 1990s and early 2000s) sharp rises again in the wake of the global financial crisis reveal the continuing strength of the British state.

Similar trends are visible across other policy domains. Social security spending has grown across the developed world. In 1980, OECD countries spent 14.4 percent of their GDP on public social expenditure. That figure had risen to 20.1 percent by 2018.[21] While this latter figure represents a modest easing off from the social expenditure costs seen after the GFC, it is clear that the age of neoliberalism has done little to cut into the aggregate social safety net.

Trends in red tape and regulation also suggest significant state

21 OECD. 2020. "Social Expenditure - Aggregated Data." *OECD.Stat.*

growth since the reform era. The US think tank Competitive Enterprise Institute conducts a yearly study into American regulatory growth.[22] Looking at the US Federal Register, a daily repository of all proposed Federal rules and regulation, gives a crude but nonetheless informative sense of the flow of regulatory activity. Where in the 1980s, there were a total of 529,223 pages in the Federal Register, the CEI predicts that by the end of the 2010s the number will grow to 777,464. The efforts of the Reagan administration and the Trump administration to reduce these figures, while substantial, did little to reduce the long-term growth trend. We can identify similar trends in Australia.[23] Just as concerning is the governance of this regulatory growth. The initation and implementation of regulation has been increasingly delegated to independent regulatory agencies—independent because they are formally separated from the usual lines of democratic accountability—with implications for democratic control.[24] As Stephen Bell and Andrew Hindmoor conclude, "a dramatic expansion in regulation started at precisely the moment in the early 1980s when resurgent neo-liberal parties were promising to 'roll back the frontiers of the state.'"[25]

---

22 Wayne Crews, Clyde. 2019. *Ten Thousand Commandments 2019.* Washington, DC: Competitive Enterprise Institute.

23 Berg, Chris. 2008. *The Growth of Australia's Regulatory State: Ideology, Accountability and the Mega-Regulators.* Melbourne, Australia: Institute of Public Affairs; 2018. "Regulation and Red Tape in a Small Open Economy." In *Australia's Red Tape Crisis: The Causes and Costs of Over-Regulation*, ed. Darcy WE Allen and Chris Berg. Queensland: Connor Court Publishing.

24 Wallace, Kurt. 2019. *Regulatory Dark Matter: How Unaccountable Regulators Subvert Democracy by Imposing Red Tape without Transparency.* Institute of Public Affairs, June.

25 Bell, Stephen, and Andrew Hindmoor. 2009. *Rethinking Governance: The Centrality of the State in Modern Society.* Cambridge, UK: Cambridge University Press. At p. 94.

Looking at this history of state growth (as well as some other indications of social liberalism like same-sex marriage and drug reform), it has been common among a group of "post-liberal" conservatives to ask: what has conservatism managed to conserve?[26] As one of these writers declared, the conservative movement needs to rethink "our practice of pursuing limited government by merely trimming the fringes of every new government expansion to keep the end result closer to what came before."[27] (The liberty movement is not entirely contiguous with the mainstream conservative movement, of course, even in Anglophone countries. While it is certainly true that the Anglo-American conservative tradition has a "classical liberal" heart, the two overlapping movements often diverge on social policies, immigration, and foreign policy.)

Of course the counterfactual is impossible to prove: how would modern political economy look without the efforts of the free market movement on legislative change and popular opinion? And lest we be seen as dismissive of that body of work, it is absolutely not our intention. Each of us have spent our career in precisely these same disputes, with (we hope) some reasonable success, slowing and even preventing particular regulatory growth and tax increases in Australia through direct engagement with public discourse and the political system. The free market movement has much to be proud of, even if long term aggregate trends are hard to dislodge.

Indeed, it is not at all clear how we ought to assess the success or failure of a political movement in an environment where there are

---

26 French, David. 2019. "Before Trump, What Did Conservatism Conserve?" *National Review*, March 8; Rothman, Noah. 2016. "What Conservatism Has Conserved." *Commentary*, October 18.

27 Cochran, Matthew. 2015. "Conservatism Is Obsolete." *The Federalist*, March 9.

equally passionate activists and political movements that support the growth of the state. It turns out that politics in a democratic system is hard, victories are never certain, and once won, always need to be defended. But perhaps we are looking for victories in the wrong places.

The economist Peter Boettke has explored at length the competing strategic visions of Friedrich Hayek and Milton Friedman.[28] The divide is relevant here. Both Hayek and Friedman were both extraordinarily gifted economists and passionate liberals—they shared a belief in the liberal order in which individuals with diverse preferences would be free to pursue their own ideas of the good life under the rule of law. Each won a Nobel for their economic contributions (Hayek in 1974, Friedman in 1976), and each had a career trajectory that shifted from their earlier focus on technical economics (Hayek in Austrian business cycle and price theory, Friedman in monetary theory) to exploring the philosophy and policy of liberalism in later life. But they had strikingly different ideas about the proper role of the scholarly economist with implications for the pursuit of liberty today.

Milton Friedman was the archetypal public intellectual who sought to persuade the public about the virtues of liberal economics. In a classic statement of the relationship between economics-as-advocacy and economics-as-analysis, Friedman wrote:

> I venture the judgment, however, that currently in the Western world, and especially in the United States, differences about economic policy among disinterested citizens derive predominantly from different predictions about the economic

28 Boettke, Peter J. 2019. "The Role of the Economist in a Free Society." *Coordination Problem*, August 7.

> consequences of taking action – differences that in principle can be eliminated by the progress of positive economics – rather than from fundamental differences in basic values, differences about which men can ultimately only fight.[29]

For Friedman, in this passage, there was relative agreement about the ends of legislation and public policy—the values that to be maximised—and only disagreement about the means. The role of the public-facing economist was to suggest the means by which they could be best achieved. We all agree that it would be better to have more wealth, rather than less; longer, healthier lives; more freedom to choose. The economist's job is to educate the public about the best ways to get to those goals; how to avoid pitfalls and unintended consequences.

That approach made Friedman one of the most important public intellectuals of the twentieth century. And it is by and large the approach adopted by most of the institutions of the liberty movement today. We explain the virtues of the liberal order to the public—or to an elite audience, who then have to explain it to the public in turn—as they relate to achieving generally agreed principles. This is even the case when it is obvious that the general public has wildly divergent views about the ultimate ends: liberty (for instance) or equality, nationalism or cosmopolitanism. Writing for the Cato Institute, the philosopher David Schmidtz described his public analytical framework as "I'm not a utilitarian, but I play one on TV."[30] The liberty movement frames policy

---

29 Friedman, Milton. 1953. "The Methodology of Positive Economics." In *Essays in Positive Economics*, edited by Milton Friedman. Chicago, IL: University of Chicago Press.

30 Schmidtz, David. 2006. "I'm Not a Utilitarian, but I Play One on TV." *Cato Unbound*, March 20.

dilemmas as how to maximise individual well-being across the social spectrum, even if its activists might be more driven by natural rights reasoning about the inherent moral virtue of liberal policy positions. This we have inherited, more than anything else, from the example of Milton Friedman.

But as Boettke points out, many of Friedman's colleagues disagreed with his argumentative approach. George Stigler, who shared Friedman's liberal worldview, distinguished sharply between the role of the disinterested scientist and the advocate, arguing in a letter to Friedman that "if a pure scientist – one believing only demonstrated things – is asked his opinion on policy, he must decline to answer – and listen to his intellectual inferiors give advice on policy. Hence the role of the pure scientist is terribly painful to assume in economics."[31] James Buchanan provided a slightly more optimistic, albeit similarly discomforting, guide to policy advice: the task of the economist is "diagnosing social situations and presenting to the choosing individuals a set of possible changes." The economist's task is complete "when he has shown the parties concerned that there exist mutual gains 'from trade.'"[32] It is up to society, then, to make the policy choices once faced with the alternatives.

It was left to Friedrich Hayek to give the definitive statement of this latter position. Hayek was not at all a disengaged scholar. His 1944 book *The Road to Serfdom*—a polemic against totalitarianism as much as a statement of liberal and economic principles—made him a global

31 Cited in: Boettke, Peter J. "The Role of the Economist in a Free Society." *Address given at the Mont Pelerin Society regional meeting in Dallas, Texas*, May 17-19, 2019.

32 Buchanan, James M. 2014. *Fiscal Theory and Political Economy: Selected Essays.* Chapel Hill, North Carolina: UNC Press Books. At p. 112.

public intellectual, especially in an abridged form published by Readers Digest. Yet Hayek was uncomfortable with the commentary and activism practiced by Friedman. In his Nobel Prize lecture, delivered in 1974, he offered a broadside against the casual certainties of "scientism" in economics. What he described as the pretense of knowledge was the confidence of economists that an irreducibly complex economy can be modelled, tamed, and then reshaped. Economics had attempted to achieve the standing of the physical sciences, but by doing so it had claimed insight into an inherently complex system that it did not have. This scientistic economics led to bad policy—Hayek noted that in the monetary inflation crisis of the 1970s, economists were being called on to solve a problem that economists had largely caused—but the certainty also eroded the value of economics as a field of study.

So here we have a sharp divide between two visions of the role of liberal intellectual. In his 1988 book *The Fatal Conceit*, Hayek cast economics along these lines: "The curious task of economics is to demonstrate to men how little they really know about what they imagine they can design."[33] Against this, the liberty movement offers advocacy: the modus operandi of Milton Friedman. But too often the liberty movement can show the confidence of the planner, asserting that bad policies cost X dollars over Y years, and that any given policy change or legislative tweak will benefit the economy by another Z dollars.

This specificity is partly because media outlets and activists are always looking for well-evidenced data that will break through otherwise commodity opinion. But it has also rightly encouraged cynicism. When a proposed reform comes up against the irreducible

---

33 Hayek, Friedrich A. 2011. *The Fatal Conceit: The Errors of Socialism.* Chicago, Illinois: University of Chicago Press. p. 76.

complexity of the economy, or the complexity of the political trade-offs, or has consequences unintended by its designers, and the dramatic gains are not immediately evident, that cynicism is completely understandable. Think of the bold claims made by the original supporters of charter schools, financial sector reform, or energy and train privatisation. Free market reform proposals are as vulnerable to the Hayekian critique as the proposals of the socialist planner.

Hayek was focused on identifying the institutional framework that would maximise individual liberty and economic growth. Friedman saw himself more as a participant within the institutional framework, taking the political system and legal environment as given and seeking to convince fellow participants to adopt liberal policy choices. The liberty movement is likewise. It is unusual, for instance, for free market think tanks to call for significant constitutional change. At the meta-political level, the liberty movement tends to favour the status quo. There are some exceptions, of course. Support for Brexit by the Institute of Economic Affairs in the United Kingdom was arguably a proposed constitutional-level change. Support for democratic plebiscites on controversial social matters such as same-sex marriage within otherwise representative democracies might be another. But for the most part the liberty movement acts within the political system as it is given, and at its best, uses its knowledge about how to convince the public and elites of pro-liberty policy reform.

But as the liberty movement remains focused on acting within national institutional frameworks, those frameworks are being disrupted by digital technology. The digital revolution has come in three stages. The first stage was the use of the computer for computation—to work faster with larger numbers and larger datasets than was possible by manual calculation. The second stage was communication—the invention of the internet. The third is the digital institutional revolution. This revolution began with the rise of platform business models—eBay, Amazon Marketplace, Facebook, Uber, AirBnB—that facilitate economic matching and exchange between strangers. More recently, blockchains and cryptocurrencies now offer an institutional infrastructure for the internet, providing a way to ensure parties to an exchange can trust each other.

It is in this third stage that the possibilities for liberty are most evident. Since the turn of the twenty-first century we have had sudden and dramatic changes in the shape of global capitalism, as new institutional technologies provide competing services against the structures of the state. They offer new mechanisms for the private provision of public goods and goods which until now most people had believed had to be provided by the government. They offer new ways of funding innovation, resolving disputes, organising and enforcing intellectual property, resolving the coincidence of wants, developing new types of social organisation and insurance, funding and selecting public goods and structuring democratic decision-making. They offer the potential to fix some of the thorniest problems in state creation, such as institutional failure and corruption.

As Jeffrey Tucker argues, new technologies open a new space for voluntary communities to form that expand our social, economic and political freedoms.[34] Technologies offer a new agenda for the liberty movement. As these new private orders become more important, the creation and development of those orders materially increases individual human liberty. Peter Leeson and Peter Boettke distinguish between two tiers of entrepreneurship: productive tier entrepreneurship

---

34 Tucker, Jeffrey. 2015. *Bit by Bit: How P2P is Freeing the World.* US: Liberty.me.

and protective tier entrepreneurship. Productive tier entrepreneurship is what we think of when we think of entrepreneurial activities: the creation of new innovations and new businesses that seek to meet anticipated consumer demand. Protective tier entrepreneurship are activities that seek to protect property rights. We see this sort of protective-tier entrepreneurship in environments where governments are weak or corrupt: the private sector steps in to enforce contracts and create communities of governance. The institutional technologies described in this book are a hybrid of the two forms of entrepreneurship. They are created in the same way as most productive tier innovations—products and services built (typically) in the developed world that come to market packaged up for consumer use—but they function as protective tier institutions—governing property rights and enforcing contracts. These institutional level, protective tier innovations provide a new guide for the liberty movement. Freedom does not only have to be fought for. Freedom can now be built.

In this book we spell out the new strategy for liberty. The first half is structured around specific technological changes—cryptocurrencies, blockchains, artificial intelligence, and privacy-preserving cryptography. We begin by exploring the revolutionary nature of cryptocurrencies. Censorship-proof, decentralised digital currencies, beginning with Satoshi Nakamoto's Bitcoin, offer an unparalleled challenge to the power of the state. Throughout the twentieth century, governments have come to rely on their control of the money supply. They produce fiat currency, they spend that fiat currency, and they insist that taxes are paid with that fiat currency. The sudden rise in non-fiat currencies—first Bitcoin, then second and third generation cryptocurrencies—has the potential to break that relationship down. We explore this new world of digital currency competition.

Underpinning digital currencies is a fundamental innovation in ledgers: blockchain technology. Blockchain technology competes with other ledger technologies to ensure trust in shared data, including governments and firms. Together with smart contracts—algorithmic contracts that self-execute on digital platforms—blockchains enable new types of economic and social organisation to emerge. Contractual agreements can now be enforced without relying on the legal systems of the state. We predict a movement away from trust provided through governments and large corporations towards trust in more decentralised blockchain networks—a process we call de-hierarchialisation.

But what does this movement away from hierarchy mean for liberty? First, it makes many of the current rationales for government intervention redundant or obsolete. Government interventions, such as antitrust or labor laws, are often justified by the need to curtail the power of large hierarchical corporations. Fewer hierarchies brings many areas of government intrusion into question. Second, blockchains and smart contracts enable new types of organisation enforced through code that can compete with government. We explore some of these mechanisms including dominant assurance contracts, personal equity swaps and endogenous digital property rights. These structures become possible in a world of blockchain-based decentralised infrastructure, and offer a new way to out-compete government for the provision of goods and services, including welfare and education funding.

We next look at what artificial intelligence (AI) and machine learning mean for liberty. These technologies have become the archetype of technological fear. Many studies look at the concerns scholars and policymakers have about the possibility of AI acting in its own interest—a so-called Skynet scenario from the *Terminator* movie series. Despite these fears, we argue that superintelligence offers an opportunity for individual liberty. Our autonomy can be augmented by these "prediction machines," helping us to make better individual choices. We argue that the response to fears about AI should be responded to through a process of *adversarial liberty*: we can protect against the challenges that AI presents through more AI competition.

The latter half of the book turns to key themes and topical issues for the liberty movement: free speech, privacy and institutional change. The rule systems that we live by—including governments—are technologies too and should be subject to competitive pressure. We investigate how entrepreneurs can propel innovation in the fundamental structure of the state—both in the developed and developing world—through experiments such as special jurisdictions and new models of democracy.

Freedom of speech is one of our most fundamental liberties, dating back at least to Ancient Greece and developing in its modern form in the battles for religious toleration. But with the internet and the explosion of global communication the speech environment is radically, and suddenly, different. We look at how technology can defeat online censorship in unfree countries—using China's Great Firewall of China as the paradigmatic example of a modern censorship regime. Decentralised censorship-resistant technologies also provide opportunities in comparatively free countries, whether censorship is at the behest of large digital platforms or regulators.

Technologies raise fraught questions for privacy. The rise of the data economy has opened vast possibilities for privacy to be violated. At the same time, state surveillance is allowing governments to pry on the affairs of its citizens. How can privacy be recaptured in the digital age? New technologies allow us to exclude not just personal communications from the eyes of the state, but economic relationships as well, with radical implications for taxation and regulation.

We explore the decoupling of privacy from secrecy and the advances of zero knowledge proofs. These technologies enable us to prove something—for example, that we are of a certain age or that we have some category medical condition—in a more granulated but private way.

Institutional competition between jurisdictions not only gives us the fundamental right to choose the rules we live under, but also helps to keep governments accountable and responsive. Because of the territorial monopoly on government and the entrenched interests of the political system, institutional competition is difficult. We argue that new technologies can facilitate institutional entrepreneurship by facilitating new special jurisdictions. There's hope for more polycentric private governance of our lives. Many challenges in lower level competing jurisdictions, however, are problems of governance and collective choice. We introduce how cryptodemocratic governance—the use of blockchain and smart contracts as a foundation to create property rights in votes—might radically expand our democratic choices. Individuals could be freer to delegate, de-compose and exchange their votes, creating more polycentric and emergent democratic structures. Cryptodemocracies facilitate a new wave of competition and choice across and within borders.

We then turn to the problem of institutional change in the developing world. One of the fundamental problems facing people in developing economies is failing governments and poor-quality institutions. Historically, attempts to fix these poor institutions have focused on change through the political system through education, or other interventionist approaches dreamt up by bureaucrats and development professionals. We argue that new technologies enable a new process of institutional change. Today entrepreneurs can layer new rules on top of existing ones within the same geographical area: breaking the relationship between institutions and geography. We explore what this process of institutional layering might look like, and how entrepreneurs with new technologies provide the opportunity for a much more decentralised, polycentric and private process of economic development. We emphasise the potential for smart contracts to create layered legal systems that compete with territorial corrupt courts.

We conclude the book with a call to arms. The new technologies of freedom exist, but they need to be developed and adopted. This means the liberty movement needs to be more entrepreneurial. We need to shift focus to the new technologies of freedom we have described—blockchains, smart contracts, artificial intelligence, privacy technologies—and focus not on shifting policy through politics, but on expanding our freedoms and prosperity by creating freedom-enhancing platforms.

02.

CRYPTOCURRENCIES AND THE STATE

Money has a bad reputation. The love of money, we are told, in the Bible, is the root of all evil. Condemnation of the love of money is found in both the Old and the New Testaments. Yet in Ayn Rand's classic work *Atlas Shrugged*, Francisco d'Anconia provides a powerful defence of money.

> "So you think that money is the root of all evil?" said Francisco d'Anconia. "Have you ever asked what is the root of money? Money is a tool of exchange, which can't exist unless there are goods produced and men able to produce them. Money is the material shape of the principle that men who wish to deal with one another must deal by trade and give value for value. Money is not the tool of the moochers, who claim your product by

> tears, or of the looters, who take it from you by force. Money is made possible only by the men who produce. Is this what you consider evil?"[1]

In a productive well-functioning economy money should be celebrated. Money is a valuable tool to facilitate trade. Money is liberating.

But at the same time, money as we experience in modern economies is a fundamental technology of state power. Governments produce our physical currency, stamp or print it with political rulers and national heroes. Through their central banks, they decide how much money is created and how slowly or rapidly its purchasing power will decline. They prop up or restrain private sector providers of money in the form of credit (that is, private banks).

Having produced the national currency, governments then insist that we pay our taxes in that currency. The power to choose which form of money taxes are collected in is one of the ways governments enforce their sovereignty. Then, once they have collected the taxes they spend that money (again, in the national currency) to hand out welfare, pay for healthcare and hospitals, public schools, infrastructure and the wages of bureaucrats and military personnel.

This tripartite relationship is an ancient one. The fact that governments print the money, tax citizens in that money and spend in that money, is a relationship that gives them absolute control over the economic life of the citizens they govern—whether or not governments choose to fully exploit that possibility. Historically, the tighter the hold that governments have had over the provision of money, the tighter a hold they have had over their citizen's lives. The move to fiat

---

1 Rand, Ayn. 2005. *Atlas Shrugged.* Penguin Books.

currency—that is, currency that is backed by no assets other than the reputation of the government that issued it—facilitated the growth of the modern state.[2]

So money is both a fundamental part of the economy, and a tool for modern state power. Wouldn't it be good if we could get the economic benefits of money, without empowering the state at the same time? This is the radical promise of cryptocurrencies—a mechanism for breaking the historical relationship between money and the overweening state. But to understand the radical opportunities for liberty offered by cryptocurrencies (and other private digital monies), we need to first understand why the government is so involved in money in the first place.

THE IDEA OF MONEY

Money is not those coloured pieces of paper (or increasingly these days polymer) that many people still carry around in their purses or wallets. Money is also not the electronic accounts that denote money that are traded in foreign exchange markets.[3]

Money is an idea. Money is a social construct. The historian-philosopher Yuval Harari has described money as being "a system of

2 Tanzi, Vito, and Ludger Schuknecht. 2000. *Public Spending in the 20th Century: A Global Perspective.* Cambridge, Massachusetts: Cambridge University Press. The classic statement of this thesis for Ancient Rome is: Crawford, Michael. 1970. "Money and Exchange in the Roman World." *The Journal of Roman Studies* 60 (1): 40-48. See also: Howgego, Christopher J. 1992. "The Supply and Use of Money in the Roman World 200 Bc to Ad 300." *Journal of Roman Studies* 82: pp. 1-31; Howgego, Christopher J. 1990. "Why Did Ancient States Strike Coins?" *The Numismatic Chronicle* 150: pp. 1-25.

3 This section is derived from an extended discussion in: Berg, Chris, Sinclair Davidson, and Jason Potts. 2019. *Understanding the Blockchain Economy: An Introduction to Institutional Cryptoeconomics.* Cheltenham, UK: Edward Elgar Publishing.

mutual trust, and not just any system of mutual trust: *money is the most universal and most efficient system of mutual trust ever devised.*"[4] Money is the product of human ingenuity in solving a complex problem—how to trade.

Money is a tool of exchange. These days we use the more formal expression "medium of exchange." In the absence of money we would live in a world of far less trade, leading to far less wealth. A barter economy—that is, an economy without money—is inefficient compared to a world with money. Inefficiency here means there are many trades that should occur and would improve our wellbeing but simply do not occur. The existence of money is a testament to human genius at solving problems.

At different times and places money has been all sorts of different commodities, from seashells to cigarettes to gold. Anything that two individuals are happy to use as money can be used as money. But a good money has to have a very specific quality: it has to be scarce. In *The Restaurant at the End of the Universe* Douglas Adams has his protagonist meet people who had chosen leaves as their money.[5] Money was literally growing on trees. Unsurprisingly, everyone was picking the leaves as quickly as they could and they had run into a "small" inflation problem.

The explanation for why any particular item emerges as a medium of exchange depends upon the other two important characteristics of

---

4 Noah Harari, Yuval. 2011. *Sapiens: A Brief History of Humankind*. Kindle Edition ed., Random House.

5 Adams, Douglas. 1980. *The Restaurant at the End of the Universe*. London, UK: Pan Books.

money: money as a unit of account, and money as a store of value.[6] When bartering you would need to be knowledgeable about all the goods and services that are available for trade, and all the possible counter-parties for a third party trade. That is a lot of detail and information to process at once. Barter exchanges are complex, involving multiple transactions at once to satisfy multiple demands. An illustration of barter complexity comes from an inscription in Cambodia before the Angkor empire, which lacked a money economy:

> And Poñ Chāñ delivered the ricefield of the Poñ which Poñ Matiśakti … delivered in reimbursement to the people of the Young God (Vraḥ Kanmeṅ), who asked in addition for 4 yau of double garments, as payment of the tax (of these people).[7]

In an economy that has money, however, all the individual needs to know is what good or service they wish to buy or sell, and what medium of exchange their counterparty would like to use. Decision making has just become a lot simpler.

The burden of exchange is reduced even further if the medium of exchange also acts as a unit of account—that is, if prices are announced (denominated) in the same terms as the medium of exchange. At this

---

6 Karl Brunner and Allan Meltzer provide an institutional analysis to answer this question, focusing on one dimension of transaction costs—information costs—which informs our argument here. See: Brunner, Karl, and Allan H. Meltzer. 1971. "The Uses of Money: Money in the Theory of an Exchange Economy." *American Economic Review* 61 (5): 784-805.

7 Cited in: Lustig, Eileen. 2009. "Money Doesn't Make the World Go Round: Angkor's Non-Monetisation." In *Research in Economic Anthropology* 29: "*Economic Development, Integration, and Morality in Asia and the Americas.*" ed. Donald C. Wood. Bingley, UK: Emerald Group Publishing.

point exchange rates between all goods and services can be measured in terms of the medium of exchange and traded in terms of the medium of exchange. The effect of this is to dramatically lower the amount of cognitive effort we have to expend to make exchanges.

Finally, we need our money to also serve as a store of value—to be ready for the next exchange when we need it. Salt losing its saltiness is a serious problem when salt is used as money. Money losing its purchasing power—inflation—is also a serious problem. Ideally money as a store of value will maintain its value over time. A dollar at the start of the decade ought to have the same economic value at the end of the decade.

In other words, changes in the price of goods and services should reflect the changes in either supply or demand for those goods and services, and not changes in the value of the money. A good medium of exchange has a stable value—if the value were expected to fall then few individuals would want to acquire and hold it and if it was expected to rise then few individuals would want to exchange it rather than hoard it. The good that simultaneously best meets those three criteria—a medium of exchange that is scarce without having a use value, which can act as a unit of account, and is a store of value—will emerge as "money" in the economy.[8]

Historically most money has been commodity based—most recently

---

8 Kocherlakota, Narayana R. 1998. "Money Is Memory." *Journal of Economic Theory* 81(2): pp. 232-251; Kocherlakota, Narayana R. 2002. "Money: What's the Question and Why Should We Care About the Answer?" *American Economic Review* 92 (2): pp. 58-61; Kocherlakota, Narayana R. 2002. "The Two-Money Theorem." *International Economic Review* 43 (2): pp. 333-346.

with gold and silver being used as "money."[9] Commodities, however, are expensive to store, difficult to transport, and tend to have use value beyond simple exchange value. They have also, often, had investment value beyond being a simple store of value. A major innovation was the substitution of a paper claim representing the underlying commodity.

Since the collapse of the gold standard and the subsequent Bretton-Woods system, however, "money" itself has come to be paper (or plastic). Money is money because the government declares it to be money. A lot of people, surprisingly but mistakenly, still think that money is backed by gold. There have been many instances in the world where fiat currency has lost its value and has been displaced by other currencies (usually the US dollar, itself a fiat currency). Yet despite these incidents, fiat currency as an institution is remarkably successful. Fiat continues to be accepted in most countries, most of the time, for most transactions, despite consistent inflation, banking crises, fiscal irresponsibility on behalf of governments, and so on.

The dominance of government-backed money is both a result of legislation and technological constraints. Governments protect their currency by enforcing legal tender laws, and by requiring taxes to be paid in the domestic currency. But Friedrich Hayek argued that the absence of market competition in the supply of money has produced money that is low quality—that is, it is inflationary and insecure, as

9 Jones, Robert A. 1976. "The Origin and Development of Media of Exchange." *Journal of Political Economy* 84 (4): pp. 757-776.

well as being technologically stagnant.[10]

Nonetheless casual observation would tend to suggest that governments are better providers of money than the private sector. After all there appears to be little demand for private money. This implies that there is some sort of market failure or inefficiency that government resolves by creating and enforcing the use of its own currency. But historical patterns suggest another rationale for money creation: governments control the supply of money in order to maximise revenue to itself.[11] Even before fiat and paper-based currencies came into existence governments manipulated the stock of money in pursuit of political goals: for instance, by holding monopolies over the extraction of the metals that constituted money, or by debasing the money stock by clipping or reforging coins. There is an intrinsic relationship between a states' taxing power, money creation power, and expenditure power. As that relationship has tightened since the end of the nineteenth century, so too have the size of governments.

The twentieth century was a period of great upheaval. In economic terms it was marked by financial instability and inflation. In political terms it was a century of warfare and revolution. These events were not unrelated to each other. By contrast the latter half of the nineteenth century seemed to be a very different era. Unsurprisingly there is a lot

---

10 Hayek, Friedrich A. 1976. *Denationalisation of Money: The Argument Refined.* London, UKL: Institute of Economic Affairs. Hayek's views on money evolved throughout his career, see: Boettke, Peter J. 2018. *F. A. Hayek: Economics, Political Economy and Social Philosophy*. London, UK: Palgrave MacMillan. pp. 67-72.

11 Selgin, George, and Lawrence H White. 1999. "A Fiscal Theory of Government's Role in Money." *Economic Inquiry* 37 (1): pp. 154-165; Ferguson, Niall. 2002. *The Cash Nexus: Money and Power in the Modern World, 1700-2000.* New York, New York: Basic Books.

of nostalgia—that pervades to this day—for the financial arrangements of that era. Ludwig von Mises, for example, writes in glowing terms of the gold standard:

> The gold standard was the world standard of the age of capitalism, increasing welfare, liberty, and democracy, both political and economic. In the eyes of the free traders its main eminence was precisely the fact that it was an international standard as required by international trade and the transactions of the international money and capital market. It was the medium of exchange by means of which Western industrialism and Western capital had borne Western civilization into the remotest parts of the earth's surface, everywhere destroying the fetters of age-old prejudices and superstitions, sowing the seeds of new life and new well-being, freeing minds and souls, and creating riches unheard of before. It accompanied the triumphal unprecedented progress of Western liberalism ready to unite all nations into a community of free nations peacefully cooperating with one another.[12]

Under the gold standard national currencies were convertible into gold on demand at a fixed rate, and then linked to each internationally by those fixed rates of exchange. The monetary historian Barry Eichengreen has argued that two features of the gold standard were vital to its continued success: credibility and cooperation.[13] By credibility

12 Mises, Ludwig von. 1949. *Human Action: A Treatise on Economics.* New York: Fox & Wilkes.

13 Eichengreen, Barry J. 1995. *Golden Fetters: The Gold Standard and the Great Depression, 1919-1939.* Oxford, UK: Oxford University Press.

Eichengreen means the confidence the public invested in government commitment to the policy. The public, apparently, had no doubt as to the priority attached by government to the continued operation of the gold standard. Central banks and governments also cooperated with each other to ensure the gold standard continued to operate at an international level by reducing instability.

Eichengreen, however, also points to some other characteristics and features of economic management at the time. The authorities did not pursue activist fiscal or monetary policy in an attempt to smooth the business cycle or reduce unemployment. Government tended to pursue balanced budgets.

The economist Ronald McKinnon argues that the most important feature of the gold standard was the "restoration rule."[14] This was the requirement that if a government did suspend convertibility between currency and gold (that is, temporarily prevent citizens from exchanging their paper currencies for the underlying commodity at the stated price) that convertibility must be restored as soon as possible, and restored at the original price. But this rule failed after all governments "temporarily" suspended convertibility during the First World War.

After the Great Depression of the 1930s was followed by the Second World War, the idea of a return to the gold standard was completely discredited by the Keynesian economists and politicians who now dominated the mainstream of economics. McKinnon explains:

> The abortive British attempt to re-establish an international gold standard from 1925 to 1931 was widely seen as having

---

14 McKinnon, Ronald I. 1996. *The Rules of the Game: International Money and Exchange Rates.* Cambridge, Massachusetts: MIT Press.

> aggravated the Great Depression. Anticipating Britain's return to her pre-war parity, the pound appreciated about 10 percent real (15 percent nominal) vis-a-vis the dollar from 1924 to 1925. Keynes guessed that this left sterling 10 percent overvalued, and that the British policy of tight money necessary to maintain this external parity was responsible for the industrial depression which Britain suffered in the remainder of the 1920s.[15]

Support for the gold standard collapsed after an abortive effort to apply the restoration rule in the 1920s imposed huge costs on the British economy and furthermore on the global economy. Ludwig von Mises, however, has a different explanation.

> The gold standard did not collapse. Governments abolished it in order to pave the way for inflation. The whole grim apparatus of oppression and coercion—policemen, customs guards, penal courts, prisons, in some countries even executioners—had to be put into action in order to destroy the gold standard. Solemn pledges were broken, retroactive laws were promulgated, provisions of constitutions and bills of rights were openly defied. And hosts of servile writers praised what the governments had done and hailed the dawn of the fiat-money millennium.[16]

At face value these two views do seem to be at odds with each other. Yet it is entirely possible that an attempted return to the gold standard

---

15 Ibid., at p. 39.

16 Mises, Ludwig von. 1953. *The Theory of Money and Credit.* Auburn, Alabama: Yale University Press. p. 420.

was a failure and politicians benefited from pursuing unconstrained and irresponsible policies in the absence of a gold standard.

In Ludwig von Mises' mind the inability of government to produce inflation under the gold standard was its major strength. The inability to pursue independent monetary and fiscal policy was a positive and not a negative. As he wrote, "The gold standard makes the determination of money's purchasing power independent of the changing ambitions and doctrines of political parties and pressure groups. This is not a defect of the gold standard; it is its main excellence."[17]

The twentieth century—following the collapse of the gold standard—saw high and variable rates of inflation. Even now prevailing economic theory and policy suggests that "low" rates of inflation are desirable. This leads to the situation whereby the central banks of the world attempt to engineer "low" rates of inflation. These central banks are the only organisations in the world that deliberately set out to lower the quality of their product offering and this is widely considered to be "a good thing."

Could we return to the gold standard? Friedrich Hayek argued that (emphasis original): "I still believe that, *so long as the management of money is in the hands of government*, the gold standard with all its imperfections is the only tolerable safe system."[18] Yet he did not think that a return to the gold standard was a practical proposition. Hayek gave two reasons: first, the gold standard was an international standard and international coordination would be required to reintroduce it;

---

17 Mises, Ludwig von. 1998. *Human Action: A Treatise on Economics*. Auburn, Alabama: Yale University Press. p. 471.

18 Hayek, Friedrich A. 1978. *Denationalisation of Money: The Argument Refined*. London, UK: Institute of Economic Affairs.

and second, the gold standard relied on the "mystique of gold" and "the general belief that to be driven off the gold standard was a major calamity and a national disgrace." Ultimately, Hayek did not believe the postwar political class had the integrity for that sort of global coordination and promise.

Hayek, however, proposed two alternatives to a gold standard. The first proposal—a commodity reserve currency—was made in the 1940s.[19] This proposal was a variation on a gold standard. It was designed to deal with one of the valid criticisms of the gold standard—that the supply of gold varied at random:

> The basic idea is that currency should be issued solely in exchange against a fixed combination of warehouse warrants for a number of storable raw commodities and be redeemable in the same "commodity unit." For example, £100, instead of being defined as so-and-so many ounces of gold, would be defined as so much wheat, *plus* so much sugar, *plus* so much copper, *plus* so much rubber, etc. Since money would be issued only against the complete collection of all the raw commodities in their proper physical quantities … and since money would also be redeemable in the same manner, the aggregate price of this collection of commodities would be fixed, but only the aggregate price and not the price of any one of them.[20]

19 Hayek, Friedrich A. 1948. *Individualism and Economic Order*. Chicago, Illinois: University of Chicago Press.

20 Ibid., at p. 212.

This proposal would operate much like a gold standard but with a basket of commodities replacing the gold. The challenge to this proposal is that, as both Mises and Hayek indicated, criticisms of the gold standard were not just technical but philosophical. Politicians did not want to be constrained. Simply meeting the criticism that gold supplies varied would not be enough to return the world to a system of sound money.

In the 1970s Hayek wrote a series of papers for the London-based Institute of Economic Affairs that proposed a choice in currency use, or the denationalisation of money.[21] Here Hayek was proposing to break the monopoly on issuing money.

> But why should we not let people choose freely what money they want to use? By 'people' I mean the individuals who ought to have the right to decide whether they want to buy or sell for francs, pounds, dollars, D-marks, or ounces of gold. I have no objection to governments issuing money, but I believe their claim to a monopoly, or their power to limit the kinds of money in which contracts may be concluded within their territory, or to determine the rates at which monies can be exchanged, to be wholly harmful.
>
> At this moment it seems that the best thing we could wish governments to do is for, say, all the members of the European Economic Community, or, better still, all the governments of the Atlantic Community, to bind themselves mutually not to place

---

21 Hayek, Friedrich A. 1976. *Denationalisation of Money: The Argument Refined*; *Choice in Currency: A Way to Stop Inflation* Sussex, UK: Institute of Economic Affairs.

> any restrictions on the free use within their territories of one another's – or any other – currencies, including their purchase and sale at any price the parties decide upon, or on their use as accounting units in which to keep books. This, and not a utopian European Monetary Unit, seems to me now both the practicable and the desirable arrangement to aim at. To make the scheme effective it would be important, for reasons I state later, also to provide that banks in one country be free to establish branches in any of the others.[22]

This idea is not nearly so radical today as it was in the mid-1970s when Hayek first proposed it. It is important to remember that the 1970s were characterised by exchange controls and severely repressed financial markets. Choice in currency is the idea that individuals should be able to transact in any currency or commodity that they chose.

> There could be no more effective check against the abuse of money by the government than if people were free to refuse any money they distrusted and to prefer money in which they had confidence.[23]

By exposing national currency to competition, governments would have to behave responsibly and maintain the value of their currency. Under such an arrangement, "those countries trusted to pursue a responsible monetary policy would tend to displace gradually those of a less

---

22 Hayek, Friedrich A. 1976. *Choice in Currency: A Way to Stop Inflation.* Kent, UK: Tonbridge Printers. p. 121.

23 Ibid., at p. 122.

reliable character."[24] Hayek did propose that various banks or other institutions issue their own currencies and that these currencies be allowed to trade alongside all other currencies. He also suggested that the notion of legal tender be abandoned, except that if the government were to issue its own currency that it should specify what currency be accepted for tax purposes, the settlement of debt, and the payment of torts.

The current domestic and international financial monetary system does not immediately resemble what Hayek called for in his proposal. Yet the monetary system does have remarkable similarities to Hayek's proposals. Governments continue to issue their own currency, but most financial institutions issue their own credit cards. People can, in many economies, hold a credit card from any bank in the world. Individuals can own bank accounts anywhere in the world—often denominated in (almost) any currency. Currencies do compete against each other in international markets, and in some countries the US dollar has displaced the local currency as the currency of choice. Exchange controls have been lifted in many parts of the world, and the control of money is largely beyond public control. Individuals can choose to contract in any currency, yet in most advanced economies are happy to use the local currency. As Hayek indicated, "unless the national government all too badly mismanaged the currency it issued, it would probably continue to be used in everyday retail transactions."[25]

The world has largely arrived at this point but not through conscious planning or even following Hayek's advice. Rather, the world arrived at this point largely through a system of trial and error. To be clear, individuals in advanced economies are usually free to organise their

---

24 Ibid., p. 123.

25 Ibid.

finances, and their bank accounts as they see fit. The greatest danger to this monetary structure is not financial repression *per se*, but rather so-called national security laws, and especially anti-liberal money laundering laws and tax harmonisation policies that have become commonplace.

THE CRYPTOCURRENCY SHOCK

Ludwig von Mises has a throw-away line at the end of his discussion of the gold standard that is somewhat prescient:

> It may happen one day that technology will discover a method of enlarging the supply of gold at such a low cost that gold will become useless for the monetary service. Then people will have to replace the gold standard by another standard. It is futile to bother today about the way in which this problem will be solved. We do not know anything about the conditions under which the decision will have to be made.[26]

Technology is yet to devise a means to convert lead to gold. Technology, however, has devised a mechanism to enlarge the supply of a close substitute for gold in a monetary system. That substitute is the manufacture of trust. Mises might not have known how technology would operate in this space. After 2009, however, we have a much clearer idea of how technology is going to revolutionise the monetary system.

The Bitcoin white paper, posted on the internet at the end of 2008, was, in retrospect a massive shock to the financial system—possibly a larger shock than the financial crisis that was sweeping through the

26 Mises (1949) at p. 473.

developed world at the time. But Bitcoin did not emerge in a vacuum. It was the first successful manifestation of a dream that dated back to the 1980s—a non-government, non-corporate native digital currency that could both facilitate transactions on the internet and undermine the political power of fiat currency.

There were many earlier attempts at building native digital currency. The most technically successful was ecash, an electronic payments system developed by the cryptographer David Chaum and released by Chaum's company DigiCash, founded in 1990. Ecash consisted of digital redeemable tokens that were cryptographically secured. One key feature of ecash was its built-in privacy protection: a consumer's bank was not able to trace individual payments back to merchants, preventing the bank from building up a spending profile on their account-holders.

An early experiment with a native digital currency was also conducted by PayPal. Years later one PayPal founder described their intention in the late 1990s to build a "global currency that was independent of interference by these, you know, corrupt cartels of banks and governments that were debasing their currencies."[27] The banking giant Citibank also piloted an e-cash system in 1997 and 2001.[28] Nonetheless, early digital currency innovations in the 1990s were commercial failures, or were replaced by more modest ambitions of offering a payment gateway between internet merchants and credit cards. As late

27 Berman, Ana. 2019. "Paypal Originally Aimed to Create Global Currency Similar to Crypto, Co-Founder Admits." *Cointelegraph*, February 1.

28 Halaburda, Hanna, and Miklos Sarvary. 2016. *Beyond Bitcoin: The Economics of Digital Currencies*. Hampshire, UK: Palgrave Macmillan. p. 113.

as 2006, it seemed that digital currencies had little commercial appeal.[29]

In retrospect however, the core weakness of these systems was their reliance on the established financial system. Every electronic currency needs to surmount what has come to be known as the "double spending problem." The double spending problem describes the technical challenge of ensuring that individual units of digital currency cannot be duplicated. One of the great features of the digital world is that electronic information can be copied endlessly without degrading. Each copy is an exact replica of its original. This is however a big problem for digital currencies, which would be useless if they could be copied, and "double spent." In the world of physical currency this is called "counterfeiting."

The digital currencies of the 1990s relied on the use of a centralised server or relationship with an established financial institution to manage the issuance and enforcement of double spending. This left companies like DigiCash vulnerable to the strategic decisions of their commercial counterparts, which was ultimately one of the reasons that Chaum's project was terminated in 1997. But more fundamentally, pre-Bitcoin digital currencies were appendages to the existing financial system, and hence highly exposed to the predations of regulators and financial competitors.

The fundamental innovation of the pseudonymous Satoshi Nakamoto's Bitcoin protocol was a mechanism to solve the double spending problem without relying on existing financial or government

---

29 White, Lawrence H. 2007. "Payments System Innovations in the United States since 1945 and Their Implications for Monetary Policy." In *Institutional Change in the Payments System and Monetary Policy*, edited by Stefan W. Schmitz and Geoffrey Wood. Routledge.

providers.[30] From a technical perspective, there is little in Bitcoin that was truly new in 2008. Like almost every major innovation, Bitcoin is cobbled together from a collection of previous innovations. Some of these were decades old when Satoshi adopted them. Public key cryptography, the system of separating a public address (key) from the private key that allows users to access information at that address, was released in 1976.[31] The Merkle trees that give each Bitcoin block its structure were developed in 1979.[32] The mechanisms that underpin the Bitcoin proof-of-work scheme (where miners compete to solve a computationally difficult puzzle) had been developed in various forms by 1997.[33] And peer-to-peer networking had gained prominence by the end of the 1990s when it was widely used for shared distribution of large files (and, indeed, for copyright infringement).

Satoshi brought these innovations together in a way that removed the need for the centralised authority that had plagued the previous attempts at digital currency. Rather than relying on a financial institution or other server to prevent double spending, Bitcoin is structured in a way that the users of the protocol itself come to a consensus over which transactions are valid and how they ought to be ordered in

---

30 Nakamoto, Satoshi. 2008. "Bitcoin: A Peer-to-Peer Electronic Cash System." *Bitcoin.org*.

31 Diffie, Whitfield, and Martin Hellman. 1976. "New Directions in Cryptography." *IEEE transactions on Information Theory* 22 (6): pp. 644-654.

32 Merkle, Ralph C. 1988. "A Digital Signature Based on a Conventional Encryption Function." In *Advances in Cryptology — Crypto '87:* pp. 369-378. Berlin, Germany: Springer.

33 Back, Adam. 1997. "A Partial Hash Collision Based Postage Scheme." *Hashcash.org*, March 28; Dwork, Cynthia, and Moni Naor. 1992. "Pricing Via Processing or Combatting Junk Mail." *Paper presented at the Annual International Cryptology Conference, 1992.*

the network. Architecturally, Satoshi invented a system for coming to social consensus over a shared digital ledger. Amendments to the ledger—that is, transactions—are grouped into blocks that refer back to the previous block. The result is a financial instrument that is virtually impossible to destroy.

Satoshi did not originally describe Bitcoin's architecture as a blockchain—the structure is referred to in the document as a "timechain."[34] This little description captures something essential about Bitcoin that the more famous 'blockchain' does not. Bitcoin's security comes from the permanence of its record. Every transaction in the chain is publicly and permanently recorded so that any individual can verify that the Bitcoins have been minted correctly and that no double spends have occurred. This leaves an immutable record going back to Satoshi's first transaction on 4 January 2009.[35] The fact that this record is shared by nodes spread across the world makes it impossible to remove transactions from the full ledger.

With our colleague Jason Potts we have described the fundamental economic process behind Bitcoin (and other proof-of-work

---

34 Rana, Gaurav. 2018. "Timechain: A Decade of Misunderstanding Blockchain." *Good Audience*, October 9.

35 It is important to point out that there are circumstances where public proof of work blockchain can be censored—most famously in the case of a 50 percent attack. Such an attack involves a single agent (or group of cooperating agents) acquiring more than 50 per cent of the hashing power of the network and preventing new transactions from being recorded. This is a serious problem for some blockchains that have little hashing power dedicated to their chain, but much less a concern for large scale blockchains like Bitcoin.

blockchains) as the *industrialisation of trust*.[36] The Bitcoin protocol takes electricity, and, by requiring the miners to expend that electricity to solve the computationally difficult puzzle, converts it into an immutable, trusted ledger. It transforms expensive energy to economically valuable trust. As a technical financial innovation this is highly significant. The existing financial system employs vast resources in order to provide trust—the reputation and authority of government central banks, walls, vaults and security guards to protect cash and gold, and digital security systems to protect digital assets. Bitcoin offers a new, remarkable way to provide trust in the stability and soundness of a digital currency.

Although it is easiest to describe it as a "new form of money"—and we will use this language throughout the book—it is important to point out that Bitcoin isn't precisely "money." On the three standard criteria discussed earlier in this chapter, Bitcoin doesn't quite stack up. It is only used by a few merchants, so it is not a full-blown medium of exchange as we usually understand it. As a unit of account, few goods or services are denominated in Bitcoin (with the notable exception of other cryptocurrencies). As a store of value, the price of Bitcoin has fluctuated wildly against national currencies, although the value of a single Bitcoin keeps its value at one Bitcoin (subject only to the fixed release of new coins in the mining process that has been specified in the Bitcoin protocol). These characteristics may change in due course. The volatility may reduce, and adoption may increase. But for now

---

36 Berg, Chris, Sinclair Davidson, and Jason Potts. 2020. "Proof of Work as a Three-Sided Market." *Frontiers in Blockchain*, January 31; Berg, Chris, Sinclair Davidson, and Jason Potts. 2019. *Understanding the Blockchain Economy: An Introduction to Institutional Cryptoeconomics*. Cheltenham, UK: Edward Elgar Publishing.

Bitcoin isn't money. Indeed, in the original Bitcoin white paper Satoshi described it primarily as a payments system.

Instead, think of Bitcoin as a new form of financial asset that has many desirable properties of money but at the same time many of the desirable properties of other types of financial instruments. The Nobel-winning economist Oliver Williamson once described a hypothetical financial instrument that combined the best features of both debt (which offers investors firm rules and consequences when debt-holders fail to service the debt) and equity (which offers investors strong discretionary control over management) that he called "dequity."[37] Bitcoin and other cryptocurrency tokens are a form of dequity. The fact that Bitcoin is programmable—that is, self-executing smart contracts can be written into the tokens themselves—gives token holders the tools to be simultaneously equity-holders and impose fixed rules financial relationships.[38]

In other words, to point out that Bitcoin does not fully display the features we would expect from money is not to denigrate it—rather, fitting it into existing economic categories is hard because it is a genuinely significant innovation. And, in common with other significant innovations, the Bitcoin white paper has launched a sustained wave of entrepreneurial creativity. Other innovators have taken Bitcoin, modified and extended it to improve its performance, or applied its best features to new domains and industries unforeseen by Satoshi.

In the next chapter we discuss in some detail how the blockchain

---

37 Williamson, Oliver E. 1988. "Corporate Finance and Corporate Governance." *The Journal of Finance* 43 (3): pp. 567-591.

38 With Jason Potts we explore this in: Berg, Davidson, and Potts, *Understanding the Blockchain Economy: An Introduction to Institutional Cryptoeconomics.*

architecture offers more than just money or money-like substitutes. We now have blockchains that offer new functions, such as a greater ability to code smart contracts, that have built in systems of governance (so that decisions about changes to the underlying protocol can be formalised and agreed to), that have alternative consensus mechanisms (to increase speed or decrease energy use), that increase the privacy of transactions, or that specialise in particular applications (like interbank transfers, or supply chains, or prediction markets, or blockchain-based organisations). Not all of these can be called "blockchains" in that they arrange transactions into blocks and cryptographically chain them together—more generally these are distributed ledger technologies.

Many of these distributed ledgers feature their own specific cryptocurrencies that can be traded as tokens. So even before we get that massive variety of applications and the challenge to centralised hierarchies that blockchain represents, we need to emphasise just how much of a revolution against the most fundamental technology of modern state power that cryptocurrencies—the first use case of blockchain—represents.

The world has 130 independent national currencies. As of December 2019 the cryptocurrency tracking site currently CoinMarketCap lists just under 5,000 cryptocurrencies. Of course, not all of those projects are actively traded or were even designed for general use. And of the few hundred cryptocurrency projects currently active, only a fraction of those have the sort of adoption that could be seen as even a small challenge to the world of national currencies. Yet the existence of these competitors in currency, now and in the future, is a profound change.

The control of government over currency is one of the most persistent features of state power. While there have been some moments in history where currencies have been provided on relatively market-based principles—such as the free banking eras of Scotland and Australia, for example—no ancient or modern money system has been unencumbered by state force. The economic function of money is to resolve the double coincidence of wants. But its political function is to broadcast the power of the government and provide resources for political action. At least some of the earliest coins we have from the kingdom of Lydia in 7th century Asia Minor, where the descendants of modern coinage was invented, are stamped with the symbol of the Lydian royal household—a lion.[39] The heads of kings and emperors on coins throughout history were not merely stamps of legitimacy, they were propaganda.

The industrial revolution saw the full capture of money-making by the state. It used its monopolisation of violence to ensure that it could monopolise the currency. Having monopolised the money used by their citizens, the size of government was able to expand dramatically, so that by the end of the twentieth century around a third of the economy of developed countries was being directly controlled by the state.

In that context, the cryptocurrency revolution has already occurred. We are just waiting to learn of its significance. Where once citizens in developed countries had access only to one currency—except for a small few who traded foreign currencies as assets—now they have access, or the potential to access, two, ten, or hundreds of alternative currencies. With the launch of the Bitcoin white paper, the age of currency monopolisation came to an end.

An illustration of the threat posed by competition in currency to the structures of government is neatly illustrated by the response of

---

39 Schaps, David. 2004. *The Invention of Coinage and the Monetization of Ancient Greece*. Ann Arbor, Michigan: University of Michigan Press.

governments around the world to the announcement of the non-government digital currency Libra. On 18 June 2019 the social media giant Facebook announced that it was working with a consortium of large firms (including, at launch time, PayPal, Mastercard, Visa, eBay, and Uber) to develop a large scale denationalised currency. Intentionally or not, Libra is a variation on Hayek's commodity currency—backing the issuance with a basket of national currencies. As the Libra white paper describes its structure:

> Libra is designed to be a stable digital cryptocurrency that will be fully backed by a reserve of real assets—the Libra Reserve—and supported by a competitive network of exchanges buying and selling Libra. That means anyone with Libra has a high degree of assurance they can convert their digital currency into local fiat currency based on an exchange rate, just like exchanging one currency for another when traveling. This approach is similar to how other currencies were introduced in the past: to help instill trust in a new currency and gain widespread adoption during its infancy, it was guaranteed that a country's notes could be traded in for real assets, such as gold. Instead of backing Libra with gold, though, it will be backed by a collection of low-volatility assets, such as bank deposits and short-term government securities in currencies from stable and reputable central banks.[40]

Libra is a permissioned blockchain. Where in Bitcoin (and other public blockchains) any technically adept individual can view, buy, mine and even develop applications on the Bitcoin protocol, a permissioned

---

40 The Libra white paper is available at https://libra.org/en-US/.

blockchain prevents anyone from doing some of those functions. Libra restricts the pool of miners—those who validate transactions on the network—to a group of 100 founding members nominated and approved by the consortium that established the Libra network.

This architecture is not unusual for a corporate-focused blockchain. One of the most widely used blockchains is Hyperledger, developed by IBM and the Linux Foundation, that allows businesses to build similarly permissioned systems for private use. A permissioned blockchain tends to offer features that are harder to build on public networks, such as greater privacy or speed. But the trade-off is that they are relatively less decentralised, reducing the most important characteristic benefit of Satoshi's invention: that there are no individual sources of authority that might be targeted by regulators or hostile actors.[41]

In the case of Libra, those sources of authority are the large firms that make up the consortium. On the very same day Libra was publicly announced, representatives of the United States House of Representatives Committee asked Facebook publicly to cease development of the network until hearings could be held.[42] The angst was bipartisan, and global. The French Minister of the Economy and Finance Bruno Le Maire declared that it was "out of the question" for Libra to be the equal of a "sovereign" currency.[43] A few months later Le Maire was more explicitly hostile, announcing Europe wide coordination to prevent the launch of the digital currency: "Libra is not welcome on

41 Some in the blockchain community dispute whether such a permissioned system can be called a "blockchain."

42 De, Nikhilesh. 2019. "Halt Libra? Us Lawmakers Call for Hearings on Facebook's Crypto." *Coindesk*, June 18.

43 Ibid.; Kuhn, Daniel. "Facebook's New Crypto Faces Scrutiny from European Authorities." *Coindesk*, June 18.

European soil … We will take steps with the Italians and Germans because our sovereignty is at stake."[44]

The fact that Facebook, which has been under political scrutiny for its use and protection of personal data, has led the development of Libra accounts for some of the instant and over-wrought political reaction. But that reaction also clearly illustrates how political leaders identify a large-scale cryptocurrency or non-government digital currency as a threat to state control over the economy. In an extraordinary letter to Visa, Mastercard and Stripe in October 2019, a group of US senators identified the risks that Libra poses for terrorist and criminal financing, that it posed to financial stability, and "interfering with monetary policy."[45] The letter concluded with an explicit threat: "If you take this on, you can expect a high level of scrutiny from regulators not only on Libra-related payment activities, but on all payment activities."[46]

In many ways, these government representatives are right to be concerned. If successful, Libra has the possibility to be the world's first private international reserve currency. This is no small thing. For the first time since the collapse of the Bretton Woods system there is a clear competitor to the US dollar for global dominance in the currency market. As a pseudo-commodity currency we should think of Libra as a return to a global gold standard. But rather than governments setting the rules and exchange rates, and gold being the underlying store of value there would be a private organisation setting the rules

---

44 Al Jazeera. 2019. "Paris, Rome, Berlin Preparing to Block Facebook's Libra in Europe." *Al Jazeera*, October 18.

45 Schatz, Brian, and Sherrod Brown. 2019. "Letter to Mr Patrick Collison." *Published letters, October 8.*

46 Ibid.

and a portfolio of relatively risk-free assets playing the role of gold.

Policymakers have been concerned about Libra because it is the first major cryptocurrency product that clearly threatens the currency monopoly held by modern states. But whether Libra is successful at its aims or not, we expect to see widespread experimentation by corporate players with alternative payments and currency systems. Libra is a lightning rod for attention precisely because policymakers fear it will be successful. Facebook alone has nearly two and a half billion users. Consumers tend to emphasise convenience, and Facebook is expert at providing intuitive consumer-technology interfaces. The financial system faces a future where billions of people adopt this new currency on the first day it is launched. The threat to currency control by the state—to currency "sovereignty"—is no longer a hypothetical dream of digital utopians or libertarian economists. At time of writing, it has a scheduled release date in 2020.

Of course, Libra may not be successful. The permissioned nature of the Libra network leaves it vulnerable to political predation—Visa, Mastercard, Stripe and PayPal withdrew from the consortium in response to regulatory pressure. But it points to a vision of a radically different global money and financial system of the twenty-first century. In testimony before the United States Congress in October 2019, Facebook founder Mark Zuckerberg pointed to the parallel creation of a fully digital currency by the Chinese central bank, with a similar global ambition but with clear political goals.[47] The European Central Bank has likewise declared its intention to develop a European digital

---

47 Zuckerberg, Mark. 2019. "Testimony to Congress." *Hearing Before the United States House of Representatives Committee on Financial Services*.

currency.[48] Separate to this effort, the French central bank announced in December 2019 a French digital currency, declaring that it "would provide a powerful lever for asserting our sovereignty in the face of private initiatives of the Libra type."[49]

As we enter the third decade of the twentieth century, the global monetary system is being dramatically realigned: a competitive dynamic between three forms of digital currency, each with their own regulatory, monetary and political characteristics. Digital currencies designed and implemented by central banks are being developed to compete against the first of what are likely to be many corporate digital currencies. And against these two are the looming fully decentralised cryptocurrencies, of which Bitcoin is the first but closely followed by the smart contract platform Ethereum, the settlement system XRP (run by Ripple), the stablecoin Tether, and hundreds of others.

How currency competition will play out is unclear. The key point here is not that any particular system is superior in every way. There are strong reasons to believe that cryptocurrencies, corporate digital currencies, and central bank digital currencies will co-exist in the future economy. But the fact that there is now a high degree of competition between systems and within national jurisdictions allows to foresee some obvious consequences. Hayekian currency competition is now being realised—and this competition will spur innovation, and punish poor competitors.

At the first instance, national currency systems will have to

---

48 Chong, Nick. 2019. "European Central Bank Wants to Beat Other Countries in Digital Currency Game." *Blockonomi*, December 14.

49 Aït-Kacimi, Nessim, and Raphaël Bloch. 2019. "The Banque De France Will Experiment with a Digital Euro in 2020." *Les Echos*, December 4.

compete—not simply rely on their control over the financial and payments system to keep their monopoly. Those central banks that tolerate high rates of inflation will see disintermediation. Increasingly borderless payments systems will mean that governments that pursue irresponsible fiscal policies will see even greater capital flight. Every country in the world faces policy challenges from a viable private international reserve currency. Control over the monetary system lies at the heart of the modern economy. A viable alternative to fiat currency, with international mobility, undermines both the conduct of monetary policy and fiscal policy. It disrupts too the cosy cartel between the political system and the banking system in many parts of the world.

FINANCE WITHOUT THE STATE

Free people require a free money. A medium of exchange, a unit of account, and a store of value that cannot be subverted. The history of finance is a story of control and subversion. Governments have long sought to control money as a mechanism to control the domestic economy. Since the collapse of the gold standard just over 100 years ago we have seen rampant inflation and an era of severe exchange controls that were relaxed only to be re-introduced under the guise of the war on terror and international tax cooperation. Activist fiscal and monetary policy has seen the unprecedented growth of government and a diminution of human freedom.

A non-government money that cannot be manipulated or controlled for political gain is very likely to hold its value and to be widely accepted. A money that does not rely on centralised trust, but rather on a private willingness to hold it and to trade it is very likely to promote human flourishing and expand trade.

The first attempts at a private digital money gave us Bitcoin—a proof of concept that demonstrated that a private digital money could work. The future very likely will be Hayekian—denationalised private monies that anchor to portfolios of assets (not gold nor commodities) but in principle underpinned by something of value. As Ayn Rand indicated, "Money is the material shape of the principle that men who wish to deal with one another must deal by trade and give value for value."

Fiat money survives because it works well enough. It is convenient. People are used to the idea that money comes from the government. The price we have paid for that convenience has been close to a century of inflation. Thanks to Satoshi and the innovators who have come after we are likely to see the emergence of an inflation proof, denationalised, valuable currency that is also convenient to use.

But the impact of currency competition goes even deeper than money itself. This change will disrupt many other related areas of government control, particularly through the heavily regulated finance sector. As we write this book the area of decentralised finance—known as DeFi—is rapidly expanding. This sector is seeking to build new financial products—from derivatives markets to loans—on uncensorable digital infrastructure, using both cryptocurrency, blockchains and smart contracting technologies (that we turn to in the following chapter). We are entering a period of finance without the state—a world where governments lose their grip not only on money, but on the financial sector more broadly.

03.

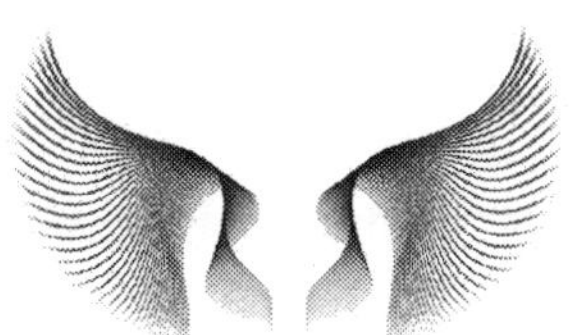

BLOCKCHAINS AND SMART CONTRACTS

The very first commercial use of the steam engine was to pump water out of mines. So the industrial revolution was born. Mechanised power gave rise to the most sustained increase in human welfare in history. The act of replacing animal and human energy (supplemented by inefficient and unreliable sources of energy such as wind and hydro) with mechanical energy allowed for the widespread adoption of mass production. The rapid decline in transportation costs and then communication costs dramatically restructured both the economy and society. Blockchain technology, or distributed ledger technology, promises a similar revolution. The industrial revolution came about because of mechanised *energy*, the blockchain revolution will come about because of mechanised *trust*. We argue in this chapter that new trust technologies will decrease our reliance on hierarchical trust from governments and corporations and that this opens the scope for more liberty.

Ludwig von Mises has defined the market system as being one of cooperation under the division of labour.[1] Markets exist to facilitate our propensity to trade. But trading itself has costs associated with it—that is, "transaction costs." Michael Munger has broken those costs into three categories.[2] What he calls triangulation costs are the costs of searching and identifying possible counter-parties as well as negotiating prices. Transfer costs are those costs involved in the actual exchange. Finally, trust costs are incurred mitigating against the risk that a counterparty will be dishonest or otherwise shirk their side of the deal.

The way we mitigate problems of trust has fundamentally shaped our world. The risk that people will act dishonestly led to elaborate legal systems, accounting principles, and corporate bureaucracies. We employ lawyers and judges, and build courthouses and prisons to ensure that we will not harm each other—that we do not steal, injure or defraud our neighbours, our clients, and strangers. We employ accountants and financial professionals to ensure that money is where we say it is—that money has not been diverted for improper uses. We hire managers to ensure that employees act in the interest of shareholders, and we hire boards to ensure that managers do the same.

These complex professions and organisational structures exist to protect us against what Oliver Williamson calls 'opportunistic'

---

1 Mises, *Human Action: A Treatise on Economics*.

2 Munger, Michael C. 2018. *Tomorrow 3.0: Transaction Costs and the Sharing Economy*. Cambridge, UK: Cambridge University Press.

behaviour, which he defines as "self-interest-seeking with guile."[3] Adverse selection and moral hazard are special cases of opportunistic behaviour. But opportunism is more than that. Opportunism goes beyond mere self-interest and includes, "the incomplete or distorted disclosure of information, especially to calculated efforts to mislead, distort, disguise, obfuscate, or otherwise confuse."[4] If we expect that our counterparties in an economic exchange are going to act opportunistically—rip us off—and that we have no tools to prevent that and they will face no penalty for doing so, then we will not trade. From this perspective, the professions and organisations that we have developed to facilitate trust have facilitated the expansion of economic activity.

Unsurprisingly, the cost of all these professions and institutions—that is, the cost of provisioning trust—is extremely high. If we look simply at the number of individuals employed in these professions in the United States, then weight that data according to how much of those professions are likely directly involved in the provision of trust, we can see that trust involves around 35 percent of all US employment.[5] If we make some heroic assumptions (such as assuming that the employment of trust providing workers in a wealthy advanced economy is representative of all economies as a whole) then $29 trillion of global GDP is involved in the provision of trust.[6] Twenty-nine trillion dollars to ensure that we do not scam or steal from each other.

---

3 Williamson, Oliver E. 1985. *The Economic Institutions of Capitalism*. New York: Free Press.

4 Ibid., p. 47.

5 Novak, Mikayla, Sinclair Davidson, and Jason Potts. 2018. "The Cost of Trust: A Pilot Study." *Journal of the British Blockchain Association* 1 (2): pp. 1-7.

6 Davidson, Sinclair, Mikayla Novak, and Jason Potts. 2018. "The $29 Trillion Cost of Trust." *Cryptoeconomics Australia*, July 24.

The cost of provisioning trust is not fixed—it is open to institutional entrepreneurship. When we get better trust technologies we can trade more. One way to reduce that cost is to develop better institutions—such as professional practices which better detect or insure against opportunistic behaviour. Another way is to develop new technologies that provide trust, freeing up human work for more productive tasks.

Satoshi Nakamoto never set out to create anything other than a decentralised digital payments system. But the blockchain architecture that Satoshi used to build this cryptocurrency has applications far beyond Bitcoin. Blockchains offer a shared record keeping system—a database, or ledger—that allows participants to come to consensus over the state of that system without the need for a central authority. Innovations in ledger technologies are important because ledgers are a fundamental economic technology—when ledgers change so too does the structure of the economy and society.

Ledgers record and verify the information jointly known about ownership, identity, relations and exchange that is necessary for economic activity to occur.[7] Ledgers confirm *who owns what*: the property title register, for instance, maps who owns land and whether that land is subject to any caveats or encumbrances. Ledgers confirm *status*: the citizenship ledger records who has the rights and is subject to obligations due to national membership. Similarly, the electoral roll ledger allows (and, in some countries, obliges) those who are on that roll a vote. Ledgers confirm business and personal *identity*. Businesses are recorded on government ledgers to track their existence and their status under tax law. Registers of births, deaths and marriages record

7 Berg, Davidson, and Potts, *Understanding the Blockchain Economy: An Introduction to Institutional Cryptoeconomics*, Chapter 4.

the existence of individuals at key moments, and use that information to confirm identities when those individuals are interacting with the world. Ledgers also record *authority*: who can validly sit in parliament, access what bank account, work with children, or enter a restricted area. Together with organisational charts that outline decision-making authority, ledgers help to *organise economic activity*. The firm itself is a ledger of relationships and identities, structuring ownership, employment, production and assets. A club is likewise a ledger, providing structure around who benefits from the club and who does not.

We share ledgers between each other to agree on the economic state of play. So when we say that the provision of trust constitutes 35 per cent of the total US workforce, we can more specifically say that a significant part of work in the economy involves ensuring that these shared ledgers reflect the real world: that our knowledge of who owns what, who is what, and who can order what is agreed to and enforced.

Satoshi's extraordinary innovation was to create an entirely new way of managing ledgers—that is, a new way to provide trust over shared information. All the accountants, managers, and regulators that govern our ledgers are themselves organised into social structures. Sometimes we buy these trust services in the marketplace, such as when we employ an auditor. Alternatively, we may rely on complex hierarchical organisations to provide the trust for us. These latter hierarchies can be states, which ensure that their ledgers are trustworthy by coercion and their monopoly of violence, or hierarchical corporations, which use their near-complete control over the labour of their employees to monitor activity within the boundary of the firm. Blockchain, by contrast, offers a non-hierarchical way to ensure trust in shared information.

In a blockchain-enabled economy some of the trust previously provided through hierarchy—either state coercion or complex corporate management—will be provided through blockchain-based decentralised networks.[8] Together with smart contracts (described in the following section) we will see a movement away from a reliance on hierarchy to a reliance on decentralised blockchain networks. We call this de-hierarchialisation—and our aim in this chapter is not only to explore this effect, but to understand the implications for liberty.

Blockchains and smart contracts are technologies of freedom. To the extent current government intervention is justified through the power of large corporate hierarchies—such as antitrust and labor regulations—a process of de-hierarchialisation takes away many of these justifications. These technologies also provide new ways to create new private coordination mechanisms, such as social insurance systems and digital property rights. The entrepreneurs that apply these trust technologies will chip away at the size of government by building a digital economy underpinned by private decentralised infrastructure.

SMART CONTRACTS AS PRIVATE LEGAL ENFORCEMENT

To understand how powerful the provision of trust by blockchain can be, we need to understand the function of smart contracts. Blockchains and smart contracts are highly complementary technologies of freedom. The Bitcoin protocol offered more than just a system to update a shared ledger. The protocol also allowed for complex code to be built into the ledger itself. That is, for transactions with conditions to be triggered and executed when certain events take place. For instance, an escrow contract built in Bitcoin could release value only

---

8 Ibid.; Allen, Darcy W.E., Alastair Berg, and Brendan Markey-Towler. 2019. "Blockchain and Supply Chains: V-Form Organisations, Value Redistributions, De-Commoditisation and Quality Proxies." *The Journal of the British Blockchain Association* 2 (1): pp. 1-8.

when certain conditions (such as the delivery of Bitcoin into a separate account) are met. In this way blockchain networks can be used as the infrastructure for "smart contracts," providing the foundation for a new way to organise more complex economic exchanges and organisations in a digital world.

Smart contracts are algorithms that trigger transfers of value in response to conditions set in code. When built into a blockchain they offer a way to make economically valuable contracts that will execute automatically, as written, everywhere there is a blockchain node—contracts which are impossible to prevent executing. Smart contracts are central to an increasingly digital economy where value is transferred over the internet. But smart contracts are not just a basic building block of the future economy, they are a powerful tool to avoid corrupt governments and overbearing regulations that might seek to prevent certain economic activity (as we will see in later chapters, blockchains and smart contracts enable truly competitive legal institutions and a vast expansion in individual choice).

Primitive versions of smart contracts pre-date blockchains. Nick Szabo described a vending machine as a smart contract: money goes in the machine, and goods (and change) are returned according to a pre-specified set of prices. As he writes, "the vending machine is a contract with bearer: anybody with coins can participate in an exchange with the vendor."[9] In the vending machine, a change to the state of the world (money goes in) acts as a trigger for the contract to execute (goods and change go out) automatically.

Are smart contracts really that smart? The founder of prominent

---

9 Szabo, Nick. 1997. "The Idea of Smart Contracts." *Nick Szabo's Papers and Concise Tutorials*.

smart contracting platform Ethereum, Vitalik Buterin, suggested that: "I quite regret adopting the term 'smart contracts.' I should have called them something more boring and technical, perhaps something like 'persistent scripts.'"[10] Indeed:

> A truly intelligent contract would take into account all the extenuating circumstances, look at the spirit of the contract and make rulings that are fair even in the most murky of circumstances. In other words, a truly smart contract would act like a really good judge. Instead, a "smart contract" in this context is not intelligent at all. It's actually very rules based and follows the rules down to a T and can't take any secondary considerations or the "spirit" of the law into account.[11]

Counter parties can be opportunistic. They can delay payment or deliver faulty goods or services. Smart contracts reduce uncertainty in exchange by reducing the chance that an agreement won't execute or be enforced.[12] They mean you no longer require confidence in your counterparty—you trust the code. But trusting hard-coded contracts rather than intermediaries or counterparties raises some other issues. In the drafting phase, for instance, counterparties can still take advantage of asymmetric information. Such drafting challenges are a problem for the legal profession, and will partly be mitigated through open standards and frameworks for

---

10 Vitalik.eth. 2018. "To be clear, at this point I quite regret adopting the term 'smart contracts'. I should have called them something more boring and technical, perhaps something like 'persistent scripts'." *Twitter*, October 13, 2018.

11 Song, Jimmy. 2018. "The Truth About Smart Contracts." *Medium,* June 11.

12 Wright, Aaron, and Primavera De Filippi. 2018. *Blockchain and the Law: The Rule of Code*. Cambridge, Massachusetts: Harvard University Press.

writing smart contracts.

In a pre-blockchain world smart contracts needed to be executed by third parties. You didn't necessarily have to trust your counterparty, but you did have to trust the intermediary who was executing the code. Since 2009 we now have a new distributed infrastructure on which these contracts can be coded and executed without relying on third parties—that technology is blockchain. Smart contracts can be coded into blockchains and execute on each full node of the chain near-simultaneously, meaning that execution and enforcement is censorship-resistant. No single authority or node can prevent a smart contract from being executed.

Recently we have seen rapid expansion in the capacity for smart contracts to be coded into blockchains.[13] While the Bitcoin blockchain enabled users and developers to implement simple escrow contracts, its programming language is rudimentary and only enables relatively simple smart contract features. The development of the Ethereum

---

13 See: Berman, Harold Joseph. 1983. *Law and Revolution.* Cambridge, Massachusetts: Harvard University Press; Trakman, Leon E. 1983. *The Law Merchant: The Evolution of Commercial Law.* Littleton, Colorado: William S. Hein & Co.; Goldenfein, Jake, and Andrea Leiter. 2018. "Legal Engineering on the Blockchain: 'Smart Contracts' as Legal Conduct." *Law and Critique* 29 (2); McKinney, Scott A., Rachel Landy, and Rachel Wilka. 2017. "Smart Contracts, Blockchain, and the Next Frontier of Transactional Law." *Washington Journal of Law, Technology Arts* 13; O'Shields, Reggie. 2017. "Smart Contracts: Legal Agreements for the Blockchain." *North Carolina Banking Institute* 21; Raskin, Max. 2017. "The Law and Legality of Smart Contracts." *Georgetown Law Technology Review* 1 (2): pp. 305-341; Ryan, Philippa. 2017. "Smart Contract Relations in E-Commerce: Legal Implications of Exchanges Conducted on the Blockchain." *Technology Innovation Management Review* 7 (10): pp. 14-21; Werbach, Kevin, and Nicolas Cornell. 2017. "Contracts Ex Machina." *Duke Law Journal* 67: pp. 313-382; Catchlove, Paul. 2017. "Smart Contracts: A New Era of Contract Use." *SSRN 3090226*; Governatori, Guido et al., 2018. "On Legal Contracts, Imperative and Declarative Smart Contracts, and Blockchain Systems." *Artificial Intelligence and Law* 26 (4): pp. 377–409.

network in 2015 included a semi-"Turing complete" programming language.[14] This advance enabled developers to build much more complex smart contracts into blockchains. Today many blockchain protocols are being developed with the capacity for more secure and simple smart contracts in mind.

But even on distributed blockchain architecture smart contracts face the problem of contract incompleteness. All possible future scenarios that need to be considered in a contract cannot be coded in a self-executing way. Figuring out which contracts can be effectively coded requires experimentation. Open-ended employment contracts, for instance, are likely to remain in the messy analogue world because they are simply too incomplete. The invention of smart contracts will incentivise the discovery of where the existing boundaries of organisations should change—for instance, some parts of employment contracts may be unbundled and coded—as well as discovering entirely new types of organisation and trade based on more complete self-executing contracts. That discovery process—of replacing hierarchal contract governance with private decentralised governance—further underscores the need for experimentation with these new technologies.

The messy smart contract adoption process will be riddled with disputes.[15] When smart contract execution goes wrong, disputes can be resolved through courts or arbitration services. For blockchain-based

---

14 Ethereum is not fully "Turing complete," in the way first described by Alan Turing, as it is limited by how many Ethereum tokens are on the network: Miller, Andrew. 2016. "Ethereum Isn't Turing Complete and It Doesn't Matter Anyway." *Youtube*, August 7.

15 For an overview see: Allen, Darcy W.E., Aaron M. Lane, and Marta Poblet. 2019. "The Governance of Blockchain Dispute Resolution." *Harvard Negotiation Law Review* 25: pp. 75-101.

smart contracts, that reversal and negotiation process isn't so simple. How will disputes over the execution of a contract be reversed between pseudonymous parties? What about different jurisdictions? How will court judgements be fed back into code and executed? For a private system of blockchain-based smart contracts to emerge and work in the real world we will require innovation in dispute resolution. The development of this new system of legal infrastructure for the internet is happening privately.[16]

Smart contracts also have security issues. Because smart contracts on public blockchains are viewable to the public, they can also be scrutinised and exploited by malicious actors. People can take advantage of bugs in the code. One of the most prominent examples of such an exploitation was of a distributed autonomous organisation (a DAO). A DAO is a series of smart contracts that relate to each other to create a governance structure. In 2016, The DAO was created as a virtual venture capital fund where people pooled money and voted on how it was invested. This process was coordinated through smart contracts on the Ethereum blockchain and was seen as a new way to democratise investment, raising over $250 million USD into the pool. But one hacker (or a group of them) found a loophole in the code, syphoning off over $50 million in cryptocurrency. This hack created a governance crisis within the Ethereum community and tested the limits of how to

16 For some discussions on the potential and the challenges of privately provided law through markets see the edited volume: Stringham, Edward Peter. 2011. *Anarchy and the Law: The Political Economy of Choice*, vol. 1. Transaction Publishers.
On some of the legal questions see also: Wright and De Filippi, *Blockchain and the Law: The Rule of Code.*

wind back an attack in a decentralised world.[17] The transition from centralised to decentralised trust won't be easy.

Smart contracts need data from the outside world to execute. Imagine a smart contract that was coded to trigger when a product was delivered to a consumer (for instance, sending a payment back to the producer of the product). How would the smart contract know that the product has reliably been delivered? The mechanisms that feed information into smart contracts that enable them to execute are known as "oracles." For smart contracts to run on deterministic blockchains they require trusted and reliable data sources. Because oracles have inaccuracies or can be tampered with, they can represent a point of weakness for smart contracts. In response to this "oracle problem"—trying to provide trusted data for smart contracts running on blockchains without integrating too much centralisation—many new oracle services are being developed, with varying degrees of decentralisation.[18] These services are necessary as a gateway for the billions of data points within an economy—such as the location of a product, or its temperature—to operate alongside self-executing smart contracts. Many of these services will be developed by the private sector as "oracle-as-a-service." For instance, prediction markets can be utilised to make oracle services more accurate and incentivise effective information feeding into smart contracts.

---

17 On the crisis see: Muhammad Izhar Mehar et al. 2019. "Understanding a Revolutionary and Flawed Grand Experiment in Blockchain: The Dao Attack." *Journal of Cases on Information Technology (JCIT)* 21 (1). See also: DuPont, Quinn. 2017. "Experiments in Algorithmic Governance: A History and Ethnography of "the Dao," a Failed Decentralized Autonomous Organization." In *Bitcoin and Beyond* edited by Malcolm Campbell-Verduyn. Routledge.

18 Note, of course, that decentralised oracle networks themselves must come to consensus.

Smart contracts are most effective when the property rights being transferred are native digital assets, such as cryptocurrency or some other unique digital property right. Enforcement is trickier when tokens represent physical assets such as the relationship between land titles and the ownership of physical land. Even if a smart contract triggers a transfer in ownership—for instance of an automobile from one party to another—that transfer also needs to happen in the analogue world:

> There is an intractable problem in linking a digital to a physical asset whether it be fruit, cars or houses at least in a decentralized context. Physical assets are regulated by the jurisdiction you happen to be in and this means they are in a sense trusting something in addition to the smart contract you've created. This means that possession in a smart contract doesn't necessarily mean possession in the real world and suffers from the same trust problem as normal contracts. A smart contract that trusts a third party removes the killer feature of trustlessness.[19]

Connecting the digital and analogue world reinforces the importance of new legal infrastructure. In the short term these new digital contracts might be sustained by operating in the "shadow of the law."[20] Enforcement, bargaining and dispute resolution might occur privately to overcome some of the costs and challenges of relying on public courts.

---

19 Song. "The Truth About Smart Contracts."

20 On private ordering in the "shadow of the law" see: Dixit, Avinash K. 2007. *Lawlessness and Economics: Alternative Modes of Governance.* Princeton, New Jersey: Princeton University Press; Mnookin, Robert H., and Lewis Kornhauser. 1979. "Bargaining in the Shadow of the Law: The Case of Divorce." *The Yale Law Journal* 88 (5): pp. 950-997.

Many of these smart contracting challenges—dispute resolution infrastructure, security, reliable oracles and physical enforcement—will be overcome in the coming decades. And as more assets are born digital—financial products, ticketing systems and intellectual property rights—they can be created, traded and enforced purely in the digital world.

Privately built and governed blockchains and smart contracts will be the foundation of the decentralised digital economy. Value can now be transferred digitally across the internet without relying on powerful hierarchies to ensure trust in transactions and to enforce contracts. Those trades will be executed and supported through a network of private digital infrastructure. For instance, blockchains and smart contracts are now being applied to track the provenance of products for consumers and governments along supply chains, the striking of trade finance and insurance contracts that execute with given events, the certification of credentials and other licenses (e.g. university degrees) that help people move across borders, the recording and transferring of land property titles in corrupt regimes, selective disclosure of sensitive information such as medical records between patients and practitioners, and as ways to organise smart cities through data markets.[21] These applications are fundamentally re-shaping the structure of global capitalism, and with it the shape of government and our liberties.

---

21 For examples of some applications see: Potts, Jason, Ellie Rennie, and Jake Goldenfein. 2017. "Blockchains and the Crypto City." *it-Information Technology* 59 (6): pp. 285-293; Ganne, Emmanuelle. 2018. "Can Blockchain Revolutionize International Trade?" World Trade Organization; Benos, Evangelos, Rod Garratt, and Pedro Gurrola-Perez. 2017. "The Economics of Distributed Ledger Technology for Securities Settlement;" Casey, Michael, et al. 2018. *The Impact of Blockchain Technology on Finance: A Catalyst for Change.* International Center for Monetary and Banking Studies (ICBM); Whitaker, Amy, and Roman Kräussl. 2018 "Blockchain, Fractional Ownership, and the Future of Creative Work." *CFS Working Paper Series.*

WHAT DE-HIERARCHIALISATION MEANS FOR THE FUTURE OF GLOBAL CAPITALISM

Much of our current corporate, political and regulatory system is structured around the need to provide trust. Any new trust technology is bound to have significant consequences on that structure. This is especially so for a trust technology that provides trust through consensus rather than hierarchically. The implications of blockchain-based infrastructure and smart contracts are profound—not just for what the future of the global economy will look like, but what this means for the relationship between governments and capitalism.

One of the major shifts in global corporate capitalism will be how our supply and value chains work. Rather than managing supply chain information in hierarchies we will increasingly rely on decentralised tracking on distributed ledgers.[22] As goods move along supply chains, information about those goods must be recorded, including their provenance, characteristics, finance—that is, their identity. Rather than recording this information in centralised ledgers, now firms can use blockchain as a shared digital infrastructure and source of truth, getting the benefits of trust in information but remaining separate legal and managerial entities. We're already seeing this new form of corporate organisation in global supply chains as ports, logistics handlers, shipping companies and large global distributors increasingly

---

22 Berg, Chris, Sinclair Davidson, and Jason Potts. *Understanding the Blockchain Economy: An Introduction to Institutional Cryptoeconomics*; Berg, Chris, Sinclair Davidson, and Jason Potts. 2018. "Outsourcing Vertical Integration: Distributed Ledgers and the V-Form Organisation." *SSRN* Paper no. 3300506; Allen, Berg, and Markey-Towler. 2019. "Blockchain and Supply Chains: V-Form Organisations, Value Redistributions, De-Commoditisation and Quality Proxies." *The JBBA* 2 (1).

adopt distributed ledgers to share information.[23]

In his famous example of the pin factory in *The Wealth of Nations*, Adam Smith identified 18 processes that workers needed to follow to produce a pin. The specialisation of labour means that these 18 processes can be divided among 18 workers. Each worker gains from the ability to focus on their individual tasks, rather than having to switch between different tasks to usher the raw materials through the production process all the way to the final pin. A specialised production process can undertake production more efficiently.

But the pin factory does not organise itself. Someone must coordinate all this work. Somehow this work must be put in order, to make sure each step in the process is of high quality. Entrepreneurs first envision a novel combination of resources to produce a profit. Managers then seek to drive efficiencies and monitor the production process. The economists Armen Alchian and Harold Demsetz explain the need for management within the firm.[24] Their argument is that firms exist to reduce the costs of monitoring team production. When many people cooperate to produce a good or service, it is difficult to identify exactly what contribution everyone has made. In a company, management provides non-market monitoring and reward systems. Management's job is to monitor whether the various tasks that need to be undertaken have been performed.

Of course, as the Marxist historian Stephen Marglin rightly points out, neither hierarchy nor the division of labour were invented

---

23 Allen, Darcy W. E. et al. 2019. "International Policy Coordination for Blockchain Supply Chains." *Asia & the Pacific Policy Studies* 6 (3).

24 Alchian, Armen A., and Harold Demsetz. 1972. "Production, Information Costs, and Economic Organization." *The American Economic Review* 62 (5).

by capitalists.[25] Rather, it is the presence of a dedicated class of management in firms that characterises a "capitalist" arrangement. Specialists in pin production cannot renegotiate contracts in real time—this would be prohibitively expensive and burdensome—thus they engage in team work. But team members may shirk their duties, free-ride off the efforts of others, or otherwise pursue their own interests while still receiving a share of total production. Hence the development of a specialised managerial function that monitors the team to ensure the profitable and productive use of joint resources.

Management is a mechanism to create trust. It is both hierarchical and centralised. One of the reasons that firms grow—that is, the reasons that firms do work themselves rather than outsource that work—is the risk that their suppliers might act opportunistically. It is hard to provide trust through market exchange. Bringing as much production as possible under the umbrella of a single management allows for better supervision of production—and has led to the growth of large companies after the industrial revolution.

Cooperation around a shared ledger provides many of the benefits of consolidation—trust—without the costs in administrative bloat and inefficiency. A blockchain enabled system that operates smart contracts will be able to provide mechanical trust in lieu of human management. Blockchains will not replace management, but they can replace some of the functions of management. Our subsequent downstream prediction is that the large multidivisional firms that characterise modern global capitalism will be less competitive against smaller firms exploiting

---

25 Marglin, Stephen A. 1974. "What Do Bosses Do? The Origins and Functions of Hierarchy in Capitalist Production." *Review of Radical Political Economics* 6 (2): pp. 60-112.

shared digital infrastructure with less administrative bloat.

Greater trust provided through blockchain networks implies a radical shift in the structure of the capitalist economy. Rather than large hierarchical organisations providing trust we will see the rise of smaller more networked firms coordinating using blockchain infrastructure. By facilitating trust in transactions and in ledgers across organisational boundaries, blockchain makes market relationships more viable, and hierarchical relationships less competitive. As some large corporate conglomerates discover that the trust in information that they had gotten from vertical integration can be provided more cheaply by shared digital infrastructure there will be a fundamental reshaping of the boundaries of firms and markets.

The twentieth century established a triumvirate relationship between employers, (unionised) labour, and labour market regulators—big business, big unions, and big government. That relationship reduced the costs of negotiating between these large sectors, facilitating the corporatist labour market regulation of the second half of the twentieth century. A combination of labour market reform and structural economic shifts (including those towards services and the digital economy) has gone a long way to undermining this dynamic, eroding the political economy foundation of regulation. De-consolidation will not be evenly distributed across the value chain, but in many cases the large firms will be replaced by networks of small and medium-sized enterprises, or SMEs, and SMEs replaced by networks of individual contractors. This powerful shift away from hierarchy as a source to trust towards more decentralised networks of trust is one of *de-hierarchialisation.*

De-hierarchialisation has enduring consequences for the structure of firms and our economy. As we dedicate fewer resources to manually maintaining trust such as through auditing—as we get better at mechanising trust—greater resources can be dedicated to innovative wealth-enhancing activities. Furthermore, as the cost of trading across jurisdictional and organisational boundaries decreases, more trade becomes possible. By lowering the costs of trade information, blockchain-based supply chain information will shift how trading networks are organised and open the scope for mutually beneficial trades. De-hierarchialisation will propel the next wave of globalisation.

But what does de-hierarchialisation mean for liberty? At the first instance attempts to squeeze these new technologies into existing regulatory frameworks will open a wide range of debates about government intervention that previously appeared settled. But there are also two deeper effects. First, many of the justifications for government interventions in the economy—namely the claim that governments are needed to control and correct for large hierarchical companies—will become redundant. Second, blockchains and smart contracts are infrastructure to solve coordination and collective action problems—such as the provision of "public goods"—that are commonly thought to be the role of the state.

BLOCKCHAINS AND THE ROLE OF GOVERNMENT

One of the key reasons for the growth of the modern state is as a counterbalance to large, non-state hierarchical organisations. At least, that is the conceit of modern economic theory. We need large political hierarchies to tame large corporate hierarchies. In a paper with Jason Potts, we map how blockchain undercuts some of the founding assumptions of mainstream economic theory.[26] For Adam Smith, the specialisation of labour drove technological change, and technological change drove an increasing specialisation of labour. But technological change and labour specialisation also produces a more complex production process. A sole producer making pins is less economically complex than a team of eighteen specialists. A more complex economy—one with longer production processes and supply chains, deeper capital investment in more specific assets (such as specialised machines)—is a more productive one. Complexity and productivity feed on each other.[27]

But as both Karl Marx and Joseph Schumpeter emphasised, economic complexity must be governed. The need for organisation induced the need for managers to coordinate that activity. And the virtuous cycle between the division of labour and technological change fed a demand for large scale capital investments. The large machines and factories needed for complex production meant that capital had to be accumulated—and a class of capitalists, who sought to invest

---

26 This follows the distinction by economist Peter Boettke between "mainstream" economic theory (contrasted with "mainline" economic theory—the tradition that threads from Adam Smith, to Friedrich Hayek, to the modern Austrians and New Institutional School). See: Boettke, Peter J. 2007. "Liberty Vs. Power in Economic Policy in the 20th and 21st Centuries." *The Journal of Private Enterprise* 22 (2): pp. 7-36; Boettke, Peter J., and Matthew D Mitchell. 2017. *Applied Mainline Economics: Bridging the Gap between Theory and Public Policy*. Arlington, Virginia: Mercatus Center at George Mason University. Berg, Chris, Sinclair Davidson, and Jason Potts. 2020. "Capitalism after Satoshi: Blockchains, dehierarchicalisation, innovation policy, and the regulatory state." *Journal of Entrepreneurship and Public Policy* 9 (2): pp. 152-164.

27 On complexity and information across economies and peoples see: Hidalgo, Cesar. 2015. *Why Information Grows: The Evolution of Order, from Atoms to Economies*. Basic Books.

their money in profitable enterprises, was born. As Marx wrote, the transformation was from "the progressive transformation of isolated processes of production carried on in accustomed ways into socially combined and scientifically managed processes of production."[28]

But the hierarchies brought by the specialisation of labour, technological change, and the need for significant capital investment and accumulation brought about problems. Marx and Schumpeter worried about the fact that the returns to scale of hierarchy risked the monopolisation of large sectors in the economy. Large hierarchies were powerful hierarchies. Marx saw capitalism as inherently tending towards monopoly. Monopolistic firms could exert power over workers (by setting wages low), consumers (by raising prices), and even the political system (by corrupting democratic decision-making).

While Marx's answer to this was a radical re-envisioning of the economic system, in the free world the solution to the problems of hierarchy came to be regulation. Economists such as A. C. Pigou (from whom we get the idea of the welfare-enhancing Pigouvian tax on externalities) and John Maynard Keynes (who argued for countercyclical government spending in times of economic distress) built up the complex regulatory system that was intended to tame the negative consequences of monopolisation and hierarchy. Corporate hierarchies would not be replaced by a single state hierarchy, as Marx predicted with his 'dictatorship of the proletariat', but they would be restrained by it. Hence we have policies like antitrust (that seeks to restrain firm growth and the exercise of monopolistic power) and labour law (that tries to prevent capitalists from pushing wages and working conditions downwards).

---

28 Marx, Karl. 1867. *Capital: A Critique of Political Economy*. vol. 1. Germany.

These policy responses have positive, negative and counterproductive effects, not least of which is their impact on innovation. The spiralling growth of the regulatory state in the twentieth century has been driven by a desire to restrain, rather than suppress, the market, and mitigate against what have been understood as its negative consequences (monopoly, excessive power). But that massive regulatory growth has reduced the dynamism of the market—the very dynamism that Adam Smith praised and underpinned wealth. Heavy regulation reduces entrepreneurial innovation. Regulation fixes businesses into established patterns, penalising entrepreneurs and innovators who try to break out of those patterns.[29] The response to this has been more public policy under the banner of "innovation policy," as governments have sought to counteract the negative effects of regulation with subsidies and incentives for innovation.[30]

In this world, policy builds upon itself. Each new regulation or subsidisation seeks to mitigate the consequences of the previous one. The government imposes prudential regulation on banks so that they don't take the risky behaviours that it's too-big-to-fail policies incentivise.[31]

---

29 Allen, Darcy W.E., and Chris Berg. 2018. *Australia's Red Tape Crisis: The Causes and Costs of over-Regulation*. Brisbane, Australia: Connor Court Books; Berg, Chris. 2008. *The Growth of Australia's Regulatory State: Ideology, Accountability and the Mega-Regulators*. Melbourne, Australia: Institute of Public Affairs.

30 Nelson, Richard R. 1959. "The Simple Economics of Basic Scientific Research." *Journal of Political Economy* 67 (3); Arrow, Kenneth J. 1962. "Economic Welfare and the Allocation of Resources for Invention." In *The Rate and Direction of Inventive Activity: Economic and Social Factors*, edited by Richard R. Nelson. Princeton, New Jersey: Princeton University Press.

31 Berg, Chris. 2016. "Safety and Soundness: An Economic History of Prudential Bank Regulation in Australia, 1893-2008." *Thesis submitted to RMIT University.*

The government provides welfare to workers that its minimum wage policies have kept out of the workforce, having already crowded out the privately-provided charities and welfare cooperatives of the nineteenth century.

The result of all this policy—complex, contradictory, and often counterproductive—is what the political scientist Steven Teles calls a "kludgeocracy," a clumsy policy system that incoherently satisfies no-one except lawyers and compliance officers.[32] Perversely, it is those same large corporate hierarchies who thrive most in a kludgeocracy, using their organisation to hire lawyers, compliance officers, and lobbyists to navigate the kludge and influence it to their ends. Rent-seeking is a natural activity for firms that will look anywhere for profitable opportunities, but it flourishes in our environment of overwhelming policy complexity and fragmentation.

Where we find ourselves today rests on a crucial institutional assumption: we need to use hierarchy to govern complex economic activity. Blockchains and smart contracts disrupt that assumption. We now have a non-hierarchical system of governance for the production processes and supply chains that have grown because of technological change and the specialisation of labour.

If the regulatory state has grown up around a desire to restrain corporate hierarchy, then how does de-hierarchialisation change that? At the first instance, it undermines one of the fundamental justifications for our kludgeocracy—if power is distributed, rather than concentrated, the need for countervailing power in the form of government action is substantially reduced. If the Marxist and Schumpeterian concerns about corporate concentration and monopolistic power are no longer

32 Teles, Steven. 2013. "Kludgeocracy in America." *National Affairs* 17 (Fall).

valid, then the question is this: are the responses to those concerns still necessary? Of course those interests that benefit from the kludge—the regulators and bureaucrats which administer it, the politicians that use it in the quest for political power, and the firms that navigate and bend it—will oppose such a wholesale rethink of the relationship between government and market. But the dilemmas that decentralised networks pose for the regulatory system are real, and they offer the possibility that markets might outflank the state.

We are already seeing these tensions around labour law and the 'gig' or sharing economy, where it is disputed whether Uber drivers should be seen as independent contractors or employees of Uber (and therefore subject to regulatory standards such as the minimum wage).[33] These relationships can be either treated as a new category of "employment"—platform engagement—or governments will have to accept that the more lightly regulated category of independent contracting will come to dominate the labour market.

But from the perspective of regulators at least there is a company called "Uber," with a CEO, and employees, and a physical office located in a physical place, which can be regulated. A decentralised Uber—that is, a protocol that allows drivers and riders to connect with each other on a blockchain or other distributed ledger—has no such target for regulators. In this world it will much be harder for governments to deliver rent to favour labour constituencies and offer social services through labour relationships. This has the potential to threaten established social safety nets and reduce labour market controls.

---

33 On regulation and the sharing economy see: Allen, Darcy W.E., and Chris Berg. 2014. "The Sharing Economy: How over-Regulation Could Destroy an Economic Revolution." *Institute of Public Affairs*, December.

Some of the very foundations of antitrust policy are also challenged by the rise of non-hierarchical economic organisation. In recent years policymakers have increasingly focused their attention on the competition effects of large digital platforms like Amazon, Google, and Facebook.[34] These platforms—known as multisided markets—are firms that allow organisationally separate users to make exchanges with each other (such as advertisers and readers, and online retailers and consumers). Competition regulators have argued that a number of these firms are monopolistic and harm consumers, and that they need to be regulated, have their mergers supervised, or even be broken up. How appropriate these recommendations are for the current generation of digital platforms is the subject of much debate. For example, the Australian competition regulator has badly mischaracterised the nature of technological change, assuming that currently dominant firms have the ability to retain dominance even in an environment where there are zero barriers to entry.[35]

Antitrust and competition policy has always seen size as a proxy for power: one or two big companies are threats to competition, but lots of small companies are a signal that the market works. Unsurprisingly when they look at digital platforms, they see a few big companies. The United States Congress can call Facebook chief executive Mark Zuckerberg to personally testify, and if they are unhappy with his answers

---

34 ACCC. 2019. "Australian Competition and Consumer Commission Digital Platforms Inquiry Final Report." *ACCC*; Digital Competition Expert Panel. 2019. "Digital Competition Expert Panel: Unlocking Digital Competition." *Government of the United Kingdom*.

35 See: Allen, Darcy W.E., et al. 2019. "Submission on the Final Report of the Australian Competition and Consumer Commission's Digital Platforms Inquiry." *Chris Berg*, September 12.

to their questions, the government can force him to split his company into smaller parts, or lower prices, or treat consumers' data differently.

But the decentralised nature of blockchain technology means that there is not necessarily a large organisation to target with antitrust action. There is no blockchain equivalent of a Mark Zuckerberg—a single agent who has managerial control over an organisation that policymakers and regulators can impose their will upon. Any regulator that sought to force changes to a distributed digital platform protocol would have to require a vast number of users, spread across the world, to make independent changes to that protocol. Just as cryptocurrencies make it virtually impossible to censor transactions, blockchain-based digital platform protocols make it virtually impossible for states to control business models. How can you regulate a monopoly when there is no monopolist?[36]

But blockchains have deeper implications for liberty than making some existing roles of the state unjustifiable. These technologies also act as infrastructure for the next wave of private governance mechanisms—people voluntarily solving coordination and collective action problems—that compete with government. Below we turn to some of the private governance mechanisms that are possible in a world of blockchains—mechanisms that directly compete with major areas of government control.

Groups of people can fail to organise themselves to act together in desirable ways because of misaligned incentives.[37] For example,

---

36 For recent theoretical work on antitrust and decentralised blockchain networks see, for instance: Schrepel, Thibault. 2020. "The Theory of Granularity: A Path for Antitrust in Blockchain Ecosystems." *SSRN* Paper No. 3519032.

37 Olson, Mancur. 2009. *The Logic of Collective Action*, vol. 124. Cambridge, Massachusetts: Harvard University Press.

"free riders" who don't contribute to a collective good mean that some goods and services may be theoretically underprovided in a free market despite the collective benefits from having those goods produced. As governments often tell us, we are too self-interested to effectively organise in groups to solve social problems, such as funding infrastructure or education—so the state steps in by taxing, spending and expanding their regulatory powers.

Private groups can collectively organise and implement mechanisms that better align incentives. Nobel Laureate Elinor Ostrom, alongside many others, demonstrated that collective action development of rules is possible.[38] That is, private collective governance is possible. Many mechanisms have been designed using insights from game theory and other fields. Recently, Eric Posner and Glen Weyl have developed a movement around their book *Radical Markets* that describes many potential new mechanisms for the provision of goods, including implementing quadratic voting in democratic governance to changing the foundations of our property rights system.[39] Posner and Weyl seek to design mechanisms to achieve their vision for "institutional arrangements that allow the fundamental principles of market allocation—free exchange disciplined by competition and open to all comers—to play out fully."[40] Complex theoretical models, computational simulations, and laboratory tests suggest that new more effective mechanisms are possible. But many of these mechanisms haven't been implemented

---

38 Ostrom, Elinor. 1990. *Governing the Commons: The Evolution of Institutions for Collective Action.* Cambridge, UK: Cambridge University Press.

39 Posner, Eric A., and E. Glen Weyl. 2018. *Radical Markets: Uprooting Capitalism and Democracy for a Just Society.* New Jersey, US: Princeton University Press.

40 Ibid., p. xvii.

because of the transaction costs of private governance, including the need for coordination between untrusted parties over long periods of time.

Blockchains and smart contracts lower the cost of developing and coordinating private governance mechanisms. Mechanisms can help to fund public infrastructure through new types of contracts, or as a foundation for new types of civil society safety nets, or to make the protection of digital property rights inherent within the infrastructure. These new mechanisms might not only help to solve the coordination or collective action problems, but also to improve the efficiency and allocation of public services. Below we turn to some examples of private governance mechanisms that become more viable in a world of decentralised blockchain infrastructure—including those that compete directly with government's capacity to intervene in the economy. These are just the beginning of a broader vision of private governance mechanisms competing with public governance.

One prominent area of government is the funding and provision of infrastructure projects such as roads, parks and bridges. Over two decades ago, economist Alex Tabarrok introduced an alternative mechanism for privately providing public goods: "dominant assurance contracts."[41] Given the technologies we have today, variations on dominant assurance contracts provide insight into how we might fund public goods.

To understand "dominant assurance contracts" we must first examine conventional "assurance contracts." Regular assurance contracts are now widely understood thanks to digital crowdfunding platforms such as Kickstarter. These platforms have raised billions of dollars

---

41 Tabarrok, Alexander. 1998. "The Private Provision of Public Goods Via Dominant Assurance Contracts." *Public Choice* 96 (3-4): pp. 345-362.

for successful projects, such as new smart watches and popular card games, using assurance contracts. On these platforms, entrepreneurs first pitch projects for which they seek funding. Private funders can then decide to pledge or invest some money towards that project. If the total amount pledged to the project doesn't meet a predetermined threshold, the project does not proceed, and the funders get their money back. Crowdfunding platforms such as Kickstarter administer this assurance contracts process to give certainty to both sides of the market.

Assurance contracts are interesting—and so-named—because they help solve the assurance problem. People contributing money can be sure that they will only pay if enough other people pay, too. But these assurance contracts don't solve the free rider problem. Why wouldn't you just wait to see if someone else funds the project? Game theoretic analysis suggests a likely equilibrium from standard assurance contracts is that people don't contribute.[42] Dominant Assurance Contracts (DACs) are interesting because they also help overcome this free rider problem. In a DAC, the entrepreneur who proposes a project agrees that if the funding threshold isn't reached then they will pay a bonus to those who pledged to contribute. This tweak shifts the incentive to contribute—if the threshold is met then the good is produced, if the threshold isn't met you get a bonus payout (paid for and agreed to by the entrepreneur at the beginning). The entrepreneur proposing the project also has skin in the game because proposing a project is no longer zero-cost, helping to reveal information about what projects should be funded.

One of the key challenges remains that experiments with such

---

42 Tabarrok, Alex. 2013. "A Test of Dominant Assurance Contracts." *Marginal Revolution*, August 29.

private mechanisms don't just need to be designed, they also need to be executed between dispersed parties who don't necessarily trust each other. Indeed, a recent DAC trial notes: "All of this is on the honour system. Please only pledge if you are serious about it."[43] Blockchain technology could be the infrastructure for experimenting with new private provision mechanisms, lowering counterparty risks and increasing the possibility of funding.[44] Indeed, as Vitalik Buterin notes:

> one can use a smart contract to implement a dominant assurance contract. Users know that they will get refunded if the contract does not hit its goal, not because they trust the operator, but because *there is no central operator they need to trust.* If the participants do not fully trust the operator to actually use the money to implement the public good, one can use a smart contract to release a small amount of money at a time, where participants can vote at any time to turn off the tap.[45]

We now have an infrastructure on which these mechanisms can be experimentally designed and built. DACs aren't perfect—there are various criticisms of their incentive structure and applicability to different types of projects—but new digital infrastructure to test and

43 Quinn, Jameson. 2013. "Matching $ Experiment: 10x Your Impact, or I'll Give You $5 Free." *Quora*, August 18.

44 See Shermin, Voshmgir. 2017. "Disrupting Governance with Blockchains and Smart Contracts." *Strategic Change* 26 (5): pp. 499-509.

45 Buterin, Vitalik. 2017. "Change the Incentives, Change the World." *Cato Unbound*, June 16.

trial with such systems is a profound advance.[46] For example, the threshold point of a DAC could be made "softer."[47] We are entering a rapid experimentation phase where entrepreneurs privately develop infrastructure to fund and provide public goods.

These same principles—of new privately built infrastructure for activities that government is thought to do—also transfer across to other major areas including charities, welfare, education and property rights protection. Many start-ups are working on applying blockchain to charities. For instance, programmable donations—and tracking the delivery of goods and services—enables donors and charities to ensure integrity in donations. This enables us not only to boost civil society philanthropy, but to pull back on a growing area of government regulation of the not-for-profit sector. AidCoin is one example of the approaches used in this space, creating services including:

> an internal exchange to convert major cryptocurrencies into AidCoin, a built-in wallet to store and donate easily, an explorer to track donations transparently, tools to connect donors with all the actors involved in the non-profit sector and templates of smart contracts to run fundraising campaigns.[48]

Even further, blockchain looks to disrupt social insurance and welfare by enabling new ways to privately insure against income loss. Government implementation of social insurance through welfare is a

46 On some criticisms see: Buterin, Vitalik. 2014. "I'm Not Understanding Why Dominant Assurance Contracts Are So Special." *Ethereum Community Forum*.

47 Quinn, "Matching $ Experiment: 10x Your Impact, or I'll Give You $5 Free."

48 AidCoin. n.d. "AidCoin Whitepaper."

public attempt to overcome the free rider problem (and the adverse selection problem) by pooling of risk. Private alternatives operating on blockchain-based infrastructure look to disrupt this role of government.

Groups of people can voluntarily agree to "income swaps," essentially creating a private social instance system. These are modern "friendly societies." Imagine coming to an agreement with five other people to swap 10 percent of your income with them every month for the next decade. Each month you would send 10 percent of your income to the five others (50 percent of your income) and in return you would receive 10 percent of the income of five others. Essentially this private nexus of income swap contracts is a private insurance system built on swapping human capital with others. Blockchains and smart contracts are critical for the success of such an income swap system to ameliorate the "monitoring and enforcement problem associated with income swaps":

> A PEC-smart contract (a smart personal equity contract, or SPEC) would be, in essence, a commitment to use a particular crypto-wallet … for all inbound income, and to link that wallet to the other counterparties in a way that cannot be unlinked except by the terms of the contract. It is in effect a voluntarily entered but hard coded mutual financial constitution among a distributed community.[49]

The idea of personal income swaps involves *swapping* income over time. But you could also *sell* shares in your future income:

---

49 Potts, Jason, John Humphreys, and Joseph Clark. 2018. "A Blockchain-Based Universal Basic Income." *Medium*, January 16.

> Someone issuing a share would sell some fraction of his future income to the market. This share would assure its owner a portion of the income received by the individual (by way of a dividend) and the right to resell that share to the market.[50]

Rather than swapping personal equity for money, your equity could be swapped for other services, including education. That is, trading a share in your income into the future for a university education today. A small charity in Cambodia, the Human Capital Project, operates by providing education using such personal equity contracts.[51] These contracts could be administered using blockchains and smart contracts, providing competition to major areas of government by unlocking global markets for personal equity.

Before blockchains digital scarcity on the internet came through centralised third-party platforms. That is, the value of digital property rights was maintained centrally—including through the power of governments. By bringing digital scarcity to the internet, blockchains can challenge government enforcement of property rights, with security provided by a decentralised network. CryptoKitties was an obscure blockchain-based game that provides insight into the power of digital scarcity. In the game, digital representations of cats are recorded as non-fungible tokens on the Ethereum blockchain. Each cat can be bought and sold using smart contracts, as well as the opportunity to breed with other cats on the platform. The platform was launched in

---

50 Clark, Joseph. 2006. "Shares in People." *Policy* 22 (1): 3.

51 See Humphreys, John. 2016. "Education in Cambodia: Rate of Return and Personal Equity Finance." *Honours thesis submitted to The University of Queensland.*

2017, and within a year over 3 million CryptoKitties had been traded—some exceeding $100,000. This is a powerful proof of concept for how digital property can be created and secured using private digital infrastructure. CryptoKitties demonstrates that blockchains and smart contracts enable us to create unique scarce tradable digital property rights. As Kate Sills notes, while CryptoKitties showed us that we can trade digital cats the implications are much broader:

> they are illustrative of more serious future endeavors. In the future, instead of representing a goofy cartoon cat, a non-fungible token might be associated with a more valuable asset, such as a company share, a concert ticket, or real property.[52]

One area that privately enforced digital property rights might impact is the creative industries and the government regulation and funding associated with it, such as innovation policy and the enforcement of intellectual property rights.[53] Further experimentations around blockchain infrastructure—such as automatically-executing royalty payments—might disrupt large areas of government control throughout the creative industries. Indeed, unique cryptographically secured assets enable:

> holders to maintain control over copyright in their cryptographic

---

52 Sills, Kate. 2018. "Code Is Capital." *libertarianism.org*, May 9.

53 See Rennie, Ellie, Jason Potts, and Ana Pochesneva. 2019. "Blockchain and the Creative Industries." In *RMIT Blockchain Innovation Hub Provocation Paper*; Potts, Jason, and Ellie Rennie. 2019. "Web3 and the Creative Industries: How Blockchains Are Reshaping Business Models." Chapter 6 in *A Research Agenda for Creative Industries*, edited by S. Cunningham. Cheltenham, UK: Edward Elgar Publishing.

> creations in gaming, collectibles, and copyright-intensive industries, to name just a few. Those industries are generally controlled by powerful, often siloed, intermediaries that restrict access, regulate transferability, and govern how or whether user-generated digital content may be controlled and exploited by the user.[54]

In this way blockchain-based digital infrastructure not only creates an internet of value, but an internet of property, putting competitive pressure on government's role as a protector of property rights. The choice of whether royalties will be paid can now be endogenously implemented into the digital property right itself by the creator. Even further, as more digital assets are created and enforced through decentralised networks on the internet, governments will struggle to encroach on property rights.

In this chapter we have seen that while the industrial revolution gave us the large modern firm and the managerial class that provided trust within it, this shift also justified labour market regulation and antitrust policy predicated on the existence of large employers and corporate hierarchies. Blockchain technology has the potential to displace large internal bureaucracies and the government interventions that seek to control them. If all blockchain technology does is simply displace some of the $29 trillion currently being spent on creating trust it will be a phenomenal innovation. We will see a profound shift across the public and private sector though de-hierarchialisation—an expanding role of markets (i.e. more horizontal networks) and a contraction in the role of hierarchy (i.e. fewer vertical networks).

---

54 Evans, Tonya M. 2019. "Cryptokitties, Cryptography, and Copyright." *American Intellectual Property Law Association Quarterly Journal* 47 (2).

The process of de-hierarchialisation we described can expand individual liberty across two main dimensions. Major areas of government growth over the past century are justified on the existence of large and powerful corporate hierarchies. Governments also rely on that hierarchical machinery to comply with those rules. De-hierarchialisation undermines not only the justification for those policy areas but also the ability for governments to enforce those policies. Even further, blockchains enable us to create new mechanisms of social and economic organisation—from funding infrastructure to private welfare systems. These technologies of freedom won't just undermine the justification and enforcement of government power, but also will directly compete with the state in providing goods and services.

04.

ARTIFICIAL INTELLIGENCE AND ADVERSARIAL LIBERTY

Artificial intelligence (AI) is a digital black box in which we pour our fears about the future of technology. It conjures up a vast array of doom-laden scenarios, from the spectre of a single hostile AI agent that might come to dominate the world, to authoritarian governments using AI to oppress their citizens, to the large firms and political campaigns using AI to predict and even control our choices.

But for all the fear and uncertainty about AI in the popular press, in many ways it is an easier technology to understand at a high level than many of the other technologies discussed in this book. To recognise what is new about modern AI is to be realistic about its prospects—and its challenges for freedom. Many existing mathematical and statistical procedures are being rebadged as AI. There is of course genuine innovation in the AI field, and AI has potentially radical implications for state power. In this chapter we explore what we mean by AI, how it came to be, and how it can be harnessed as a tool of individual rights and autonomy.

There are significant challenges that come with the development of this technology, and many policymakers have begun to call for new bureaucracies and regulators to regulate AI algorithms. But as politicians and regulators turn their attention to this extraordinary new technology, the answer is not more regulation—that is, not more state power.

Many beneficial technologies, such as the printing press, have been used by the state and other bad actors in negative ways. But as is often said by advocates of liberty, that the solution to bad speech is more speech. As we shall argue, we can tackle the harmful uses of AI with beneficial uses of artificial intelligence. Where the state undermines our liberties, we can use technology to try to reclaim them. We call this *adversarial liberty*—the defence of freedom using the same technologies that the state (and other bad actors) uses to threaten it.

DEMYSTIFYING ARTIFICIAL INTELLIGENCE

All sorts of machines have been described as "intelligent." An 1857 edition of the London-based *Gentlemen's Magazine* wrote of the arithmometer, an early tabletop calculator:

> Instead of simply reproducing the operations of man's intelligence, the arithmometer relieves that intelligence from the necessity of making the operations … It is not matter producing material effects, but matter which thinks, reflects, reasons, calculates, and executes all the most difficult and complicated

> arithmetical operations with a rapidity and infallibility which defies all the calculators in the world.[1]

The arithmometer did not live up to these high standards. A calculator cannot think, reflect, or reason; only calculate and execute. As one early definition of AI suggested, the goal of AI researchers is to develop machines that behave as though they were intelligent.[2] It is the "as though" which is important. Intelligence has long been understood as a human quality—the ability to *think*. The AI project is to design and deploy machines that simulate those human qualities.

The "ability to think" is a far too imprecise goal for practical innovation. As a consequence, many of the approaches to testing whether AI research is moving towards this goal have focused on practical comparisons between human and machine thought. The famous Turing test, described by the mathematician Alan Turing in 1950, is a game of imitation where a human interrogates a machine—through typewritten questions and answers, in his original formulation—and tries to guess whether it is human or an AI agent pretending to be human.[3] The questions used by the human operator in the Turing test are meant to be open-ended. The human could ask anything, so the computer needs to be able to answer questions from out of left field.

But the imitation of thought is not thought. As the philosopher John Searle has argued, even if a computer can be designed to mimic how

---

1 The Gentleman's Magazine. 1857. "A New Calculating Machine." *The Gentleman's Magazine*.

2 Ertel, Wolfgang, and Nathanael T. Black. 2018. *Introduction to Artificial Intelligence*. Springer International Publishing.

3 Turing, Alan M. 1950. "Computing Machinery and Intelligence." *Mind* 59 (236): 433-460.

a conscious being acts, it is a category error to describe that computer as conscious.[4] Human brains and computer "brains" are different things—the mind is not just a computer program, and the process of thought in the human mind and the process of thought in a digital, programmed environment are not the same.

Nonetheless the Turing test captured something economically critical about the development of AI—and the culture built around it. The useful question isn't whether robots can become, in a philosophical sense, humans, but whether they can be observed to be human-like, or if they can do work that is perceived by other humans to be human-like work. Writing after the Second World War, Turing limited his test to whether a computer would be perceived as human in a written conversation (when a "small-scale" computer weighed 3,000 pounds). A generalisable version of the Turing test might be that a physical computer—a robot—could convince a human that it had performed in any given task as well (or as poorly) as a human could. Regardless of whether we could describe a computer as having a mind or consciousness, it would be hard to ask for more.[5]

One side effect of this approach however is that it puts deception at the forefront of the human-AI relationship. The computer is impersonating a human, and is trying to fool humans.[6] It is unsurprising then

---

4 Searle, John R. 1980. "Minds, Brains, and Programs." *Behavioral and Brain Sciences* 3 (3): 417-457.

5 Harnad, Stevan. 2001. "What's Wrong and Right About Searle's Chinese Room Argument?" In *Essays on Searle's Chinese Room Argument* edited by M. Bishop and J. Preston. Oxford University Press.

6 Bringsjord, Selmer, Paul Bello, and David Ferrucci. 2003. "Creativity, the Turing Test, and the (Better) Lovelace Test." In *The Turing Test.* Springer. pp. 215-239.

that fears about the dangers of human-ish computers were established early. Turing was still working on wartime cryptanalysis at Bletchley Park when the science fiction author Isaac Asimov published his "laws of robotics"—rules designed to prevent robots from doing harm to their human owners.[7]

But consider the Turing test from the perspective of a computer. The computer receives a question or a statement to react to—a string of letters followed by punctuation. As it doesn't 'understand' language in the sense that we humans do, it needs a mechanistic way to interpret that string of letters and then return a plausible human-seeming response. One way to do so might be for programmers to develop a list of all possible questions, and design answers in response. Then the programmers would develop decision trees for situations where questions have follow-ups, and so on. Known as "classical AI" or just "good old-fashioned AI," these expert systems follow paths deterministically.[8] But this obviously places a massive burden on the designers—and such structured, rule-based call-and-response would be hard to describe as "intelligent."[9]

By contrast to the classical model, most modern approaches to AI focus on *learning*. Rather than programming up every possible response, the computer is taught to discover, for itself, what answers work and what do not. This is why Ajay Agrawal, Joshua Gans and Avi Goldfarb of the University of Toronto's Creative Destruction Lab

---

7 Asimov, Isaac. 1942. "Runaround." *Astounding Science-Fiction* 29 (1).

8 Named in: Haugeland, John. 1989. *Artificial Intelligence: The Very Idea*. Cambridge, Massachusetts: MIT Press.

9 Pearl, Judea. 2019. "The Limitations of Opaque Learning Machines." Chapter 2 in *Possible Minds: Twenty-Five Ways of Looking at AI* edited by John Brockman. Penguin Press.

characterise the services we call modern AI not as *intelligence*, per se, but as *prediction*.[10]

A learning computer in a Turing test tries to predict what the string of letters means to a human and which responses will be the most convincing. How does it make these predictions? Typically by analysing how humans have historically responded to those strings of letters. For example, when asked, "How are you?" the average human will respond "I'm good." So the computer predicts that the human operator in the Turing test will accept the "I'm good" as a human-like answer. Over time, the learning computer will also consider which of its predictions were successful (that is, which answers were accepted by the human operator) and which were not (in other words, answers which triggered a failure of the Turing test). Doing this requires a massive amount of data on which those predictions can be based. Each subsequent attempt at an answer provides the computer with more information. The computer trains and improves over time, making more accurate predictions—better, more convincing answers to questions through the brute force of experience.

Learning machines are everywhere: website chatbots, robotic call answering services, and the now common smart devices' virtual personal assistants like Google Home, Siri and Amazon's Alexa. AI decision-making engines power autonomous (that is, driverless) vehicles. As an autonomous vehicle travels down a street, it is using the information from those sensors and the maps to make predictions about what it "sees"—vehicles, bicyclists, pedestrians, roadworks, and

10 Agrawal, Ajay, Joshua S Gans, and Avi Goldfarb. 2018. *Prediction Machines: The Simple Economics of Artificial Intelligence* Harvard Business Review Press.

other obstacles. By observing how other users of the road act, it can make predictions about whether they will move into the autonomous vehicle's lane, or otherwise require the computer to change course or speed. The results of this massive data collection are then fed back into data that can help inform vehicles facing similar obstacles in the future—whether helping build or update a fine-grained map of the street, or understanding the behaviour of other road users.

Modern AI in this way is both incredibly powerful—a virtuous cycle of observing, acting, and learning—and seemingly mundane. Judea Pearl, one of the key architects of this approach to AI, has argued that while this approach to AI is effective, it has intrinsic limits. First, it is opaque. The nature of the training means that it is sometimes hard for external observers (users, operators, even regulators) to identify why a computer has made any given decision. In general, the more complex the data set and the more advanced the technique, the less transparent it is. Neural networks, which mimic the adaptive and dynamic structure of the human brain and can identify patterns in data that are not evident to humans, are infamously opaque.[11] As Pearl writes,

> I find many users who say that it "works well and we don't know why." Once you unleash it on large data, deep learning has its own dynamics, it does its own repair and its own optimization, and it gives you the right results most of the time. But when it doesn't, you don't have a clue about what went wrong and what should be fixed. In particular, you do not know if the fault is in

11 Knight, Will. 2017. "The Dark Secret at the Heart of AI." *MIT Technology Review*, April 11; Burrell, Jenna. 2016. "How the Machine 'Thinks': Understanding Opacity in Machine Learning Algorithms." *Big Data & Society* 3 (1): 1-12.

> the program, in the method, or because things have changed in the environment.[12]

And as Pearl points out, the statistical AI approach seems unlikely to achieve some of the more radical visions of the future of artificial intelligence. John Searle famously distinguished between two visions, which he described as *weak AI* and *strong AI.*[13] Weak AI focuses on prediction—the sort of pattern matching invaluable for autonomous vehicles, for identifying diseases in X-rays, for matching photographs to biometric data. It is in weak AI that the huge advances in machine learning and neural networks have occurred over the last two decades.

Strong AI is the AI of *I, Robot,* where not only can computers compute, but they can reason. Also known as *artificial general intelligence*, this is the project of developing truly intelligent agents out of computational components. Where the prediction-based learning can identify correlations in data, strong AI can identify causes. While there is some dispute about how human-like AI would have to be to be considered artificial general intelligence—yes, it should be able to reason, but should it also have emotions? Yes, it should be creative, but should it also have imagination?—progress towards strong, artificial general intelligence has been halting, at best.

---

12 Pearl, Judea. 2019. "The Limitations of Opaque Learning Machines." Chapter 2 in *Possible Minds: 25 Ways of Looking at AI* edited by John Brockman. Penguin Press.

13 Searle, "Minds, Brains, and Programs."

IS ARTIFICIAL INTELLIGENCE DANGEROUS?

Isaac Asimov wrote his laws of robotics in the service of science fiction, not as part of a research program into AI safety. The laws are comprehensive enough to be plausible, but ambiguous enough that he could derive the plots of stories when they went wrong:

- First: A robot may not injure a human being or, through inaction, allow a human being to come to harm.
- Second: A robot must obey orders given it by human beings except where such orders would conflict with the First Law.
- Third: A robot must protect its own existence as long as such protection does not conflict with the First or Second Law.[14]

Asimov later added a "zeroth" law—to indicate its overriding importance:

- A robot may not harm humanity, or, by inaction, allow humanity to come to harm.[15]

There are a lot of holes in these laws, which Asimov could exploit for dramatic effect: what happens if two humans give a robot contradictory orders? What does it mean to harm humanity? If murder is prevented by the First and Fourth Laws, is stealing? Is prying on humans harmful? Is ugliness harmful?[16]

That four rules seems inadequate even as a statement of principles

---

14 Asimov, "Runaround."

15 Asimov, Isaac. 1986. *Robots and Empire.* HarperCollins.

16 Turner, Jacob. 2019. *Robot Rules: Regulating Artificial Intelligence.* Palgrave Macmillan.

is unsurprising: the "don't harm others" standard that applies to human action in liberal societies is in fact made precise through hundreds of laws and thousands of interpretive legal cases, and enforced through elaborate and complex legal systems. Common law has evolved in order to capture precisely the sort of situational variation in harm.[17] As they argue, just as an economic planner lacks the information to outperform the distributed market, a legal planner lacks the information to outperform a distributed common law system that adjudicates specific cases and specific disputes. Asimov offered us both the top-down planner approach to AI safety, and, through his stories built around the Laws, a critique of how that general rule approach can go wrong.

The canonical warning of the danger of general rules in the control of AI is the Paperclip Maximiser Scenario. Consider an AI agent with a single programmed and apparently harmless mandate: to maximise the number of paperclips it produces. The AI agent need not be a strong AI with emotions or reason, just a learning algorithm focused on paperclip production. Eventually the paperclip maximiser, myopically focused on its one goal, consumes all the metal on the planet, then all the other resources, then all the matter in the universe—including humans—in pursuit of more paper clips.

In his book *Superintelligence*, Nick Bostrom offers the most serious and comprehensive exploration of the scenarios of the development of strong general AI that he describes as superintelligence.[18] A super-

---

17 Leoni, Bruno. *Freedom and the Law*. 1961. The William Volker Fund Series in the Humane Studies. Princeton, New Jersey: Van Nostrand; Hayek, Friedrich A. 1973. *Law, Legislation and Liberty,* Volume 1: Rules and Order. Chicago, Illinois: University of Chicago Press.

18 Bostrom, Nick. 2014. *Superintelligence: Paths, Dangers, Strategies.* Oxford, UK: Oxford University Press.

intelligent agent is one whose intelligence vastly outstrips that of the humans who designed it—a future in which humanity is not the most intelligent group on the planet. A superintelligent machine is a machine that can improve itself—not just learn by doing, as with today's prediction engines, but one which (given that it is more intelligent than its human designers), will be able to improve its capabilities on the margin of intelligence. It is an intelligent machine that can make itself more intelligent.[19]

Given how we as humans currently treat the less intelligent species on our planet, this superintelligence brings with it obviously existential risks for humanity. Bostrom cites the chief statistician of Alan Turing's code-breaking team that, "the first ultraintelligent machine is the last invention that man need ever make, provided that the machine is docile enough to tell us how to keep it under control."[20] There is no reason to believe that a superintelligent agent will adopt human ideas of right and wrong. And a superintelligent agent with a non-human moral compass might be motivated by the acquisition of useful resources—humans among them. This is what is typically called the AI value alignment problem: how can we get a superintelligent machine to adopt our values?[21] Asimov's Laws of Robotics are one fictional attempt at aligning values through rules.

---

19 Good, Irving John. 1966. "Speculations Concerning the First Ultraintelligent Machine." In *Advances in Computers* 6: pp. 31-38.

20 Cited in: Bostrom, Nick. 2014. *Superintelligence: Paths, Dangers, Strategies*. Oxford University Press

21 Yudkowsky, Eliezer. 2016. "The AI Alignment Problem: Why It's Hard, and Where to Start." In *Symbolic Systems Distinguished Speaker series,* Stanford University, May 5; Tallinn, Jaan. 2017. "AI and Value Alignment." In *Beneficial AI.*

Bostrom is concerned not only with the existence of this superintelligence but its path to development. A slow takeoff for superintelligence would be manageable. Humans know how to deal with risky new technologies. We can mitigate their risks, regulate their consequences, and otherwise control them, however imperfectly. But a fast takeoff—a development of self-improving superintelligence in the space of weeks, days or even hours—would be much more dangerous. The *Terminator* franchise's Skynet presents the hard take-off scenario. Skynet is a superintelligent agent that once brought online assesses humanity as a danger and responds by immediately attacking it. It then exploits its superior intelligence by improving its capabilities—most obviously in the successive *Terminator* models—faster than humans can improve theirs.

Bostrom judges a fast takeoff—the sudden appearance of a superintelligence—as more likely than a slow takeoff, with dangerous implications. The fact that we can only judge the intelligence of an AI against our understanding of human intelligence could well be blinding us to the rapid pace of its development. The task is then to focus on AI safety while it seems far-fetched, not leave it until we are being chased into a bunker by autonomous drones. Taken at its limit, Bostrom's scenario is horrifying. But one of the problems with the discussion of AI risk and AI safety thus far is its lack of economic reasoning. Just as Asimov used his Laws as a critique of top-down legal rules, the dangers of superintelligence provide a window into the value of polycentric institutional orders.

We already have a superintelligence: human civilisation. Civilisation, write Christine Peterson, Mark S. Miller and Allison Duettmann, "is vastly more intelligent than any individual, it is already composed of both human and machine intelligences, and its intelligence is already increasing at an exponentially accelerating rate."[22] This superintelligence is a non-humanly designed superintelligence. It is the superintelligence of global coordination around markets, societies, and cultures that is described so well by the mainline tradition of economic thought from Adam Smith to the latter day Austrian economists.[23] The spontaneous order of complex global economic activity is as hard to observe as any black box AI. This superintelligence however does not have a single "utility function"—it does not act with a singular purpose. Rather, it is shaped by the diverse goals and preferences of billions of individual agents.

This comparison has some significant implications for the AI safety debate. To the extent that AI can become a superintelligence, it will be the planet's *second* superintelligence—competing for the same "domination." Human civilisation and strong AI obviously differ along some critical margins—the latter is designed and unidirectional (that is, it has a goal, such as paperclip maximisation, and deploys its resource to that singular end), whereas civilisation is unplanned and multidirectional (there is no end goal, *per se*, of humanity, just the goals and desires of a billions of individuals and groups).

Bostrom's worst case scenarios are disturbing precisely because they feature a single, directed superintelligence, emerging fast enough to dominate or suppress competing superintelligence projects. Thus,

---

22 Peterson, Christine, Mark S. Miller, and Allison Duettmann. 2017. "Cyber, Nano, and AGI Risks: Decentralized Approaches to Reducing Risks." *Paper presented at the Colloquium on Catastrophic and Existential Risk, University of California, Los Angeles, United States, March 27-29.* p. 1.

23 Boettke, Peter J. 2012. *Living Economics: Yesterday, Today, and Tomorrow.* Independent Institute; Boettke, Peter J., S. Haeffele-Balch, and V.H. Storr. 2016 *Mainline Economics: Six Nobel Lectures in the Tradition of Adam Smith.* Mercatus Center, George Mason University.

Bostrom concludes, we need to be focusing on how to control that monopolistic AI. In the Peterson, Miller and Duettmann framework, to introduce control over superintelligence is to harness one super-intelligence against another. What we need is competition for superintelligences. If AI monopoly leads to danger, then AI polycentrism—competitive artificial intelligences, competing on different margins, in different jurisdictions, for different economic problems—provides a way through.

In a comprehensive treatment Eric Drexler offers an alternative vision of the development of AI (what he calls "comprehensive AI services" or CAIS) that treats the development of AI as a competitive ecosystem.[24] The focus on a sudden step-change in AI capabilities envisaged by Bostrom misses the mark. Rather, we can observe in the real-world AI development that is both incremental, competitive, and in the service of humans. As he writes, "Focusing exclusively on relatively distant prospects for full automation would distract attention from the potential impact of incremental research automation in accelerating automation itself."[25]

We are currently in the midst of the 'intelligence explosion' that has been predicted for decades, but that explosion is distributed across many fields and applied to discrete tasks, not concentrated in a singular superintelligence. Modern AI is focused on specialisation, not generalisation—we have AI focused on searching within discrete spaces for knowledge, whether that knowledge is better targeted advertising

---

24 Drexler, Eric K. 2019. "Reframing Superintelligence: Comprehensive AI Services as General Intelligence." In *Technical Report No. 2019-1*. Future of Humanity Institute, University of Oxford.

25 Ibid., p. 19.

on social media, identifying diseases on X-rays, or detecting obstacles for autonomous vehicles. Indeed, the competitive drive in AI, Drexler argues, is towards optimising for specific tasks, not developing a strong AI that could choose its own tasks and motivations. In this framework, AI checks AI. The future of superintelligence is adversarial: multiple AI systems operating in the same jurisdiction, but with diverse tasks and specialisations. This process prevents any given system from dominating:

> In the CAIS model, it is natural to produce diverse, independent, task-focused AI systems that provide adversarial services. By contrast, it has been argued that, in the classic AGI model, strong convergence (through shared knowledge, shared objectives, and strong utility optimization under shared decision theories) would render multiple agents effectively equivalent, undercutting methods that would rely on their independence. Diversity among AI systems is essential to providing independent checks, and can enable the prevention of potential collusive behaviors.[26]

Drexler's formulation of AI risks not only offers a more grounded view about how AI is developing, but also a model about how to think about AI safety. Rather than focusing on building systems of control before a superintelligence takeoff, we ought to be focusing on how to ensure AI systems are being developed competitively. Indeed, AI researchers and developers are deeply familiar with the adversarial model—there are many AI learning approaches which use an adversarial system that pits two AI systems against each other, as if a computer

26 Ibid., p. 77.

was both asking and answering the questions in a Turing test.[27] The liberty movement should also become familiar with this adversarial model, adopting adversarial liberty as a way to generate more freedom.

To see how adversarial liberty might work, we now consider one of the most highly publicised AI dangers: the creation of "deepfake" videos which have been used to convincingly superimpose the faces of famous individuals on other footage. Deepfakes represent a threat that state and non-state actors will undermine core ethical norms—even harm the very notion of truth in the public space. In the next chapter we consider the use of AI by the state to suppress freedom of speech. But looking at the deepfake problem first shows us how we can use an adversarial liberty approach to protect against harmful AI.

DEEPFAKES AND ADVERSARIAL LIBERTY

Few AI applications epitomise the concerns about reckless use of AI more than deepfakes. Deepfakes are videos generated using AI that "swaps" images of one face with another face dynamically. Deepfakes are typically generated by neural networks that identify the contours of a face on the original video and map the replacement face onto those contours. Deepfakes take advantage of two AI-related technologies: the statistical modelling of human faces, and techniques that can generate videos from still images.[28] Having been developed by the academic AI community, faceswapping launched into public consciousness when a

27 Huang, Ling et al. 2011. "Adversarial Machine Learning." *Paper presented at the Proceedings of the 4th ACM workshop on Security and artificial intelligence, Chicago, Illinois, United States, October 17-21.* See pp. 43–58.

28 For a review of these techniques see: Zakharov, Egor et al., "Few-Shot Adversarial Learning of Realistic Neural Talking Head Models," *arXiv preprint arXiv:1905.08233* (2019).

user on the online forum Reddit posting under the name "deepfakes," began superimposing the faces of famous actors onto pornography videos.[29] When Reddit pulled the deepfakes down, they migrated to more obscure corners of the internet.

Doctored videos in this way raise deep questions about privacy, sexual harassment, and, more generally the reliability of information presented to the public. After the Canadian professor Jordan Peterson discovered videos manipulating his image into singing rap songs and a website that allowed users to customise the Peterson deepfakes, he wrote:

> I can't imagine what the world will be like when we will truly be unable to distinguish the real from the unreal, or exercise any control whatsoever on what videos reveal about behaviours we never engaged in, or audio avatars broadcasting any opinion at all about anything at all.[30]

The *Washington Post* has argued deepfakes only deepen the problem of "fake news": "Trust is eroding, social media is accelerating the disintegration by allowing lies to spread at unseen speeds, and the leader of this country is joining our enemies in taking advantage of it."[31] And in June 2019 the US Congress held a hearing to investigate how the government should respond to what it saw as "a potentially

---

29 Cole, Samantha. 2017. "AI-Assisted Fake Porn Is Here and We're All Fucked." *Vice*, December 12.

30 Peterson, Jordan. 2019. "The Deepfake Artists Must Be Stopped before We No Longer Know What's Real." *National Post*, August 23.

31 The Washington Post. 2019. "Deepfakes Are Dangerous — and They Target a Huge Weakness." *The Washington Post*, June 17.

grim, 'post-truth' future."[32] There are some suggestions that deepfake techniques have been used to maliciously doctor videos of political figures with genuine political consequences, although as of writing this has not been clearly seen in any developed democracies.[33]

It is often observed that social problems rarely have technological solutions.[34] But, then again, rarely do social problems have technological causes. Doctored photographs long predate digital technologies. One widely produced portrait of Abraham Lincoln was in fact Lincoln's face superimposed on John C. Calhoun's body. Totalitarian regimes have long practised photo manipulation to remove inconvenient parts of their own history. And of course, lies have been part of the art of politics since politics began.[35] Misleading and erroneous information is a permanent feature of political debate. In the wake of the 2016 US presidential election, many media commentators fixated on the effect that "fake news" may have had on the vote—deliberately false news stories that were propagated through social media, such as Pope Francis

---

32 US House of Representatives Intelligence Committee. 2019. "House Intelligence Committee to Hold Open Hearing on Deepfakes and AI: The National Security Challenge of Artificial Intelligence, Manipulated Media, and 'Deepfakes.'" *House Intelligence Committee* news release, 7 June.

33 The major cited instance has been in Gabon, where an allegedly deepfake video is claimed to have added impetus to an attempted military *coup d'etat.* See: Breland, Ali. 2019. "The Bizarre and Terrifying Case of the "Deepfake" Video That Helped Bring an African Nation to the Brink." *Mother Jones*, March 15.

34 Morozov, Evgeny. 2013. *To Save Everything, Click Here: The Folly of Technological Solutionism.* PublicAffairs eBook.

35 Arendt, Hannah. 1972. *Crises of the Republic: Lying in Politics, Civil Disobedience on Violence, Thoughts on Politics, and Revolution.* Harcourt Brace Jovanovich; Johnstone, S. 2011. *A History of Trust in Ancient Greece.* Chicago, Illinois: University of Chicago Press.

endorsing the candidate Donald Trump for election.[36]

In that sense, the effect of deepfakes on public trust is one of degree not kind. Confirmation bias—the desire to search for and interpret new information in a way that confirms one's pre-existing beliefs—is a psychological phenomenon, not a technological one. Even for those who are panicking about the consequences of deepfakes, the most compelling concern is that they could *further* undermine trust in information, not that they undermine trust *per se*.

Ever since the fake news panic, technology companies—particularly the social media giant Facebook and search giant Google—have invested significantly into trying to detect misleading content using AI, with the ultimate goal of either flagging information as misleading or simply removing it from their platform. Manual, human-based verification of whether a story is "fake" is extremely labour intensive, and the spread of misleading information can be extremely rapid. "A lie will gallop halfway round the world before the truth has time to pull its breeches on," said President Franklin Roosevelt's chief of staff Cordell Hull.[37] Given that, automating fake news detection is extremely appealing.

There are three typical approaches to fake news detection: focusing on how the information is propagated (fake news might have particular patterns of propagation), the context in which it is shared (certain demographics seem to be more susceptible to sharing fake news than others,

---

36 This example is cited in: Allcott, Hunt, and Matthew Gentzkow. 2017. "Social Media and Fake News in the 2016 Election." *Journal of Economic Perspectives* 31 (2): 211-236. One particularly direct example of this panic is: Read, Max. 2016. "Donald Trump Won Because of Facebook." *New York*, November 9.

37 Hull, Cordell. 1948. *The Memoirs of Cordell Hull*, vol. 1. New York: Macmillan Company. p. 220.

and receive different patterns of likes and reactions), and the content of the fake news itself (focusing on linguistic styles and language choices).[38] But each of these techniques have obvious limitations. As Claire Wardle, a researcher into the mitigation of disinformation online puts it, "For machines, how do you code for 'misleading'? … Even humans struggle with defining it. Life is messy and complicated and nuanced, and AI is still a long way from understanding that."[39]

By contrast, compared against the deeper problems of political lying, deepfakes are a more straightforward technical challenge. While it might be hard for casual observers to determine whether an image or video is authentic, this is not the case for professionals. The field of digital image forensics is focused on determining the origins and veracity of digital images, beginning with traditional image forensic techniques (such as observing how shadows fall on elements within the picture) to identifying artefacts on the image from processing and compression.[40] Additional digital forensic techniques that can be applied to deepfakes involves looking for duplicated or dropped

---

38 Kai Shu et al. 2017. "Fake News Detection on Social Media: A Data Mining Perspective." *ACM SIGKDD Explorations Newsletter* 19 (1); Zhou, Xinyi, and Reza Zafarani. 2018. "Fake News: A Survey of Research, Detection Methods, and Opportunities." *arXiv preprint arXiv:1812.00315*; Monti, Federico, et al. 2019. "Fake News Detection on Social Media Using Geometric Deep Learning." *arXiv preprint arXiv:1902.06673.*

39 Strickland, Eliza. 2018. "AI-Human Partnerships Tackle 'Fake News'." *IEEE Spectrum*, August 29.

40 Bohme, R., and M Kirchner. 2013. "Counter-Forensics: Attacking Image Forensics." Chapter 12 in *Digital Image Forensics: There Is More to a Picture Than Meets the Eye*, edited by Husrev Taha Sencar and Nasir Memon. New York: Springer-Verlag; Johnson, Micah K., and Hany Farid. 2005. "Exposing Digital Forgeries by Detecting Inconsistencies in Lighting." *Paper presented at the Proceedings of the 7th Workshop on Multimedia and Security, New York, United States, August 1-2.*

frames in a video, or cloned areas of a single frame.[41]

In a high-stakes environment like a courtroom, it is worth investing in digital forensic professionals to adjudicate disputed videos.[42] But platforms and publishers face different challenges. The skilled labour needed is costly, and the distribution of fake information is fast. Here AI systems trained to identify the patterns and features of digital images and videos associated with deepfakes are a powerful tool. Researchers have developed datasets of manipulated videos and photographs to train deepfake-detecting AI agents.[43] In other words, the same AI techniques that have helped create misleading videos can be used to automate the process of detecting them. Platforms and publishers of videos can then flag videos for human review or just notify watchers that they have been detected as potentially misleading. Forensic AI counters deepfake AI.

Other mechanisms to mitigate audio-visual fraud use blockchain technology to verify the veracity of videos. Blockchains provide a public and unalterable record where audio-visual content can be registered, allowing later recipients of that content to validate the provenance of the information. To the extent that original content can be recorded on the blockchain as close to the moment of its production

---

41 Bestagini, Paolo, et al. 2013. "Local Tampering Detection in Video Sequences." *Paper presented at the 2013 IEEE 15th International Workshop on Multimedia Signal Processing (MMSP), September 30 - October 2.*

42 For a discussion on the use of contested digital video in court, see: Maras, Marie-Helen, and Alex Alexandrou. 2019. "Determining Authenticity of Video Evidence in the Age of Artificial Intelligence and in the Wake of Deepfake Videos." *The International Journal of Evidence & Proof* 23 (3): pp. 255-262.

43 Rössler, Andreas, et al. 2018. "Faceforensics: A Large-Scale Video Dataset for Forgery Detection in Human Faces." *arXiv preprint arXiv:1803.09179.*

as possible, the technology provides an infrastructure for verification.[44] An adversarial anti-deepfake AI might also be able to use that infrastructure to support its detection. Of course, blockchain is hardly a complete solution to the problem of deepfakes. But to the extent that it can be built into the production processes and workflows of content creators, it underscores the possibilities brought about by the simultaneous development of multiple general purpose technologies—not just AI and blockchain, but also the Internet of Things devices that could record information automatically, and communications networks that ensure high-bandwidth connectivity.

The underlying technology which allows people to create deepfakes can be used to protect rights, not just place them at risk. It is significant that some of the early significant research into face swapping was motivated by concerns about the protection of privacy. In a 2008 paper, a team out of Columbia University described a system of automatic face replacement in images, and described the purpose in this way:

> Advances in digital photography have made it possible to capture large collections of high-resolution images and share them on the internet. While the size and availability of these collections is leading to many exciting new applications, it is also creating new problems. One of the most important of these problems is privacy ... Identity protection by obfuscating the face regions in the acquired photographs using blurring, pixelation, or simply

---

44 For discussions on this see: Hasan, Haya R., and Khaled Salah. 2019. "Combating Deepfake Videos Using Blockchain and Smart Contracts." *IEEE Access* 7: pp. 41596-41606; Cai, Yuanfeng, and Dan Zhu. 2016. "Fraud Detections for Online Businesses: A Perspective from Blockchain Technology." *Financial Innovation* 2 (20).

> covering them with black pixels is often undesirable as it diminishes the visual appeal of the image. Since the number of images being captured is growing rapidly, any manual approach will soon be intractable. We believe that an attractive solution to the privacy problem is to remove the identities of people in photographs by automatically replacing their faces with ones from a collection of stock images.[45]

Currently most services which seek to anonymise imagery do so by blurring or pixelating identifying information (faces). When Google launched its Street View service as part of Google Maps, there was a widespread controversy when it was discovered that the camera vehicles which drove along the streets of the developed world to collect street images were also capturing the faces of pedestrians, violating a sense of privacy even for those who are travelling out of their home. Google Maps Street View now blurs out faces.[46] But these techniques have weaknesses. Blurring often fails to ensure privacy. It alters the image significantly enough to be both less valuable for other services to build on and less valuable aesthetically.[47] And researchers have demonstrated that deep learning techniques can be used to

---

45 Bitouk, Dmitri, et al. 2008. "Face Swapping: Automatically Replacing Faces in Photographs." *ACM Transactions on Graphics (TOG)* 27 (3).

46 Shankland, Stephen. 2008. "Google Begins Blurring Faces in Street View." *CNET*, May 13.

47 See for instance: Neustaedter, Carman, Saul Greenberg, and Michael Boyle. 2006. "Blur Filtration Fails to Preserve Privacy for Home-Based Video Conferencing." *ACM Transactions on Computer-Human Interaction (TOCHI)* 13 (1).

reidentify individuals who have been pixelated or blurred out.[48] Here the technology behind deepfakes can provide support for privacy. Adversarial AI offers a technology for the automated anonymisation of faces while simultaneously preserving the aesthetic value of imagery and video.[49]

Deepfakes undoubtedly present a complex new challenge—even if the problem of lying in politics is not new. But those challenges will be solved by researchers and entrepreneurs, not regulators. As harmful uses of AI are developed, we rely on other researchers and entrepreneurs to develop countermeasures—techniques that protect against, or at least mitigate, the harms. This process looks like an arms race. But an arms race is a reasonable description of the competitive process around innovation. Entrepreneurs create new products and bring them to market. Competitors respond by making better products. Indeed the metaphor of a "race" is common in the academic literature around innovation.[50] While the margin of competition is serious—individual liberties, rather than, say, the competition for formulate better toothpaste—the dynamics are not unusual.

The arms race metaphor also offers us a guide for action. Most of the commentary around deepfakes—and harmful private uses of AI more generally—has focused on whether it can be regulated by government, or even banned, in the interests of privacy and liberty. But the rapid

---

48 McPherson, Richard, Reza Shokri, and Vitaly Shmatikov. 2016. "Defeating Image Obfuscation with Deep Learning." *arXiv preprint arXiv:1609.00408*.

49 For example, see: Hukkelås, Håkon, Rudolf Mester, and Frank Lindseth. 2019. "Deepprivacy: A Generative Adversarial Network for Face Anonymization." *arXiv preprint arXiv:1909.04538*.

50 Khanna, Tarun. 1995. "Racing Behavior Technological Evolution in the High-End Computer Industry." *Research Policy* 24 (6): 933-958.

development of these technologies means that government action would be unlikely to be successful, even if it was desirable. It may be the case that social problems do not have technological solutions. But they can have entrepreneurial ones.

Artificial intelligence raises a vast number of new challenges for liberty. In this chapter we have considered the existential threat of an AI superintelligence and the more practical and immediate challenges of fake information (in the form of deepfakes). Adversarial approaches—the use of AI for liberty against the use of AI for harm—opens up new opportunities for the protection of liberty and individual rights. In the next chapter we explore how this sort of adversarial AI can be used to protect one of most important freedoms in the liberal canon: freedom of speech.

05.

FREEDOM OF SPEECH IN THE DIGITAL AGE

John Stuart Mill influentially argued that those who are most harmed by restrictions on speech are listeners, not speakers.[1] Censorship denies the audience the opportunity to test their own views against those of others. The more speech—and the more diverse that speech—the better.

It is easy to forget how radically different the information environment of the twenty-first century is compared to the past. One common, but ultimately unprovable claim, made in the 1980s was that the average weekday edition of *The New York Times* contains more information in it than a person in the seventeenth century was likely to come across in

1 Mill, John Stuart. 1859. *On Liberty*. London, UK: J. W. Parker and Son.

their lifetime.[2] The information scientists Martin Hilbert and Priscila López have tried to quantify much more clearly the sharp increase in information over the course of the twentieth and twenty-first centuries so far.[3] They find that the technological capacity to store and communicate information grew 30 percent per year between 1986 and 2014—five times the pace of economic growth. This increase has of course been dominated by digitisation. In 2014 less than half of one percent of information was stored on analogue media.

Not only is there more information, but it is more accessible than ever before. The combination of digitisation, mobile internet, and fully featured pocket computers (what we narrowly and now anachronistically call "phones") means that virtually all public information is available anywhere, at any time. In 2019, 3.2 billion people on the planet had a smartphone.[4] And unlike subscribers to *The New York Times* in the 1980s, today's consumers have an historically unparalleled ability to seek the information they want, on their terms, without the overbearing curatorial hand of an editor determining what stories are most important. Of course, previous generations could always experience this information diversity by going to a library, but the information on the internet is magnitudes greater than any library and is immediately and constantly accessible through a small screen.

---

2 Wurman, Richard Saul. 1989. *Information Anxiety.* California, United State: Doubleday.

3 Hilbert, Martin. 2012. "How Much Information Is There in the 'Information Society'?" *Significance* 9 (4): pp. 8-12; 2017. "Information Quantity." In *Encyclopedia of Big Data* edited by Laurie A. Schintler and Connie L. McNeely. Switzerland: Springer. For a survey of methods see: 2015. "A Review of Large-Scale 'How Much Information' inventories: Variations, Achievements and Challenges." *Information Research* 20 (4).

4 Newzoo. 2019. "Global Mobile Market Report." *Newzoo.*

Just as consequentially, this information revolution has broken down the distinction between those who produce information and those who consume it. The "citizen journalists" of the 2000s and the explosion of social media in the 2010s have not only undermined the power of curators and editors but created a global, virtual, and borderless public sphere, where readers and writers not just intermingle but effortlessly switch roles. As the legal scholar Eugene Volokh writes

> Speakers' desires are fairly simple: generally, they want more listeners. But listeners don't just want more speakers talking to them. Listeners want more control over their speech diet – a larger range of available speech coupled with greater ease of selecting the speech that's most useful or interesting to them.[5]

Journalists employed by traditional print mastheads release their breaking news on Twitter, and the videos and reports of ordinary citizens as they find themselves living through important moments are featured on those very same mastheads. Where decades ago media theorists and regulators feared that our information environment was shaped by a small number of media moguls, the fear is now that there is too much information for us to process—that there is, as one head of Australia's press council declared, a "cacophony" of speech, and that that extreme quantity somehow harms freedom of speech.[6] The scale of this communications revolution is now very familiar. But its

---

5 Volokh, Eugene. 1996. "Freedom of Speech in Cyberspace from the Listener's Perspective: Private Speech Restrictions, Libel, State Action, Harassment, and Sex." *The University of Chicago Legal Forum*.

6 Murphy, Katherine. 2011. "We Need to Lift Our Game, Press Watchdog Says." *The Sydney Morning Herald*, November 9.

implications are not. We have more ability to speak (and to read) than ever before, and it has materially changed the environment for one of our most fundamental liberties: freedom of speech.[7]

The idea of freedom of speech emerged in the battles over religious toleration in the early modern period. It became part of the liberal canon during the Enlightenment. Throughout the twentieth century freedom of speech chipped away at sedition laws, government secrecy, and anti-obscenity statutes. While this trajectory has not been linear (the postwar era saw the emergence of anti-'hate' speech statutes in most developed countries with the notable and remarkable exception of the United States, and national security legislation has revitalised sedition laws) it is nonetheless the case that with some important exceptions in liberal democracies the freedom to express political views has never been greater.

And yet the rapid digitisation and growth of information has moved the battlegrounds of freedom of speech into new arenas, both presenting new challenges and offering new solutions to censorship. Authoritarian states now target digital communication and use powerful new tools—like artificial intelligence—to censor information. In the free world, the increasingly central role that social media platforms like Facebook and Twitter play in public debate have led many conservatives to argue that the ability those platforms have to censor certain types of speech and delete accounts is a new threat to speech. In this chapter we tackle these new domains for free speech in turn—focusing first on the historically unparalleled Great Firewall of China, and then the question of social

---

7 Berg, Chris. 2012. *In Defence of Freedom of Speech: From Ancient Greece to Andrew Bolt.* Monographs on Western Civilisation. Melbourne, Australia: Institute of Public Affairs.

media censorship. In each, we are interested in how a new generation of technologies empowers entrepreneurs and activists to undermine attempts to prevent us from speaking to each other.

FREEDOM OF SPEECH IN CHINA

Information control has been at the centre of the Chinese Communist Party's control since its founding. As one influential Chinese journalist wrote, the CCP's political power:

> must depend on two weapons: guns and pens … the ruling party's 'two weapons theory', can almost be interpreted as: in the midst of humanity's unstoppable trend toward democracy, the CCP wants to keep its power by controlling public opinion and by violence.[8]

Mass media—newspapers, radio, television—are seen as vehicles for disseminating information from the party to the public, rather than as an institutional watchdog on the party.

When the internet came to mainland China in the mid-1990s, the Chinese state both aggressively pursued its introduction as a form of industry policy to build the digital economy, and sought to put tight controls on its use, banning content that, for instance, "harms the interest of the nation," or "spreads rumors or disturbs social order."[9] In practice, the Great Firewall of China (officially named the "Golden

---

8 Qiang, Xiao. 2004. "A Bold New Voice - Lu Yuegang's Extraordinary Open Letter to Authorities." *China Digital Times*, July 20.

9 Roberts, Margaret E. 2018. *Censored: Distraction and Diversion inside China's Great Firewall*. New Jersey, United States: Princeton University Press.

Shield Project") blocks many major foreign websites (including news and social media platforms) directing Chinese users onto websites hosted on the Chinese mainland media which are then subject to their own censorship regimes (such as blocks on keywords, such as "Tiananmen").

Techniques exist to bypass the Great Firewall. Between three and 11 percent of Chinese internet users use circumvention tools to avoid censorship.[10] For example, virtual private network services (VPNs) allow users to direct all their internet data through an accessible foreign server. But these techniques are available only to a limited few, as the process of installing a VPN cannot typically be done from within China itself, and a VPN is only as useful as its foreign server is available. The Chinese government goes through periodic crackdowns on VPN accessibility.[11] Other technologies are similarly vulnerable. One prominent anti-censorship tool is Tor (originally an acronym for "The onion router"), which sends internet traffic through a series of relays ("bridges") to obfuscate the Tor user's identity. But the Chinese state has proven adept at identifying and blocking secret bridges, meaning that Tor has only limited use in China.[12] Thus, despite these

---

10 Shen, Fei, and Zhi'an Zhang. 2018. "Do Circumvention Tools Promote Democratic Values? Exploring the Correlates of Anti-Censorship Technology Adoption in China." *Journal of Information Technology & Politics* 15 (2); Roberts, Hal, et al. "2010 Circumvention Tool Usage Report." *The Berkman Center for Internet & Society*, October.

11 Leng, Sidney, Josh Ye, and Nectar Gan. 2017. "The Who, What and Why in China's Latest VPN Crackdown." *South China Morning Post*, January 26.

12 Winter, Philipp, and Stefan Lindskog. 2012. *How the Great Firewall of China Is Blocking Tor*. USENIX-The Advanced Computing Systems Association; Dunna, Arun, Ciarán O'Brien, and Phillipa Gill. 2018. "Analyzing China's Blocking of Unpublished Tor Bridges." Paper presented at the 8th USENIX Workshop on Free and Open Communications on the Internet (FOCI).

techniques, the effect of the Great Firewall has been to ensure there are in practice two internets, each with its own online retailers, social media services, payment gateways and so forth—and the Chinese state tends to proclaim the value of censorship as a form of industry policy.[13]

No society in human history has ever managed to fully close off information to its citizenry. In the Soviet Union and other Eastern Bloc countries, clandestine *samizdat* publications were reproduced manually and distributed through dissident groups, in what has famously been described as the Soviet "pre-Gutenberg" culture.[14] Soviet dissidents used typewriters and handwritten documents to distribute their work. But modern dissidents have a much wider variety of tools.

The Chinese #MeToo movement offers a poignant example of the evasion of censorship by the new technologies of freedom. #MeToo rocketed to global attention in the West when extensive allegations of sexual harassment and assault were made against the American film producer Harvey Weinstein in 2017, and the #MeToo hashtag used on social media to indicate how widespread harassment, assault and rape were throughout society. China's #MeToo has been the target of online censors—with the original English hashtag being quickly

---

13 The Global Times. 2015. "What Impact Does the Firewall Bring to the Chinese Internet?" *The Global Times*, January 28.

14 Komaromi, Ann. 2015. *Uncensored: Samizdat Novels and the Quest for Autonomy in Soviet Dissidence.* Northwestern University Press; Skilling, Harold Gordon. 1989. *Samizdat and an Independent Society in Central and Eastern Europe.* Macmillan Press, in association with St Antony's College, Oxford.

blocked, and Chinese language replacements blocked.[15] #MeToo is seen as a form of collective political action and a potential threat to the established order.

Yet even at the height of this censorship during 2018, China's online censors were unable to prevent #MeToo accounts being shared permanently. One student of Peking University posted a detailed account of university intimidation when she and a group of other students delivered a freedom of information request about rape on campus on the Ethereum blockchain.[16] The message was encoded in a zero-dollar transaction (subject to a small transaction fee) in a data field which can be used to hold any arbitrary data. The transaction, along with its message can now be viewed by any Chinese user with access to the Ethereum blockchain or a web-based Ethereum blockchain navigator, impervious to the censorship power of the state.[17] Similarly, when the coronavirus whistleblower doctor Li Wenliang passed away in February 2020, a post appeared on the Ethereum blockchain memorialising him and criticising the actions of the Chinese government. In this way, blockchain technology provides a mechanism for dissidents to pass messages behind and through the Great Firewall.

Indeed, researchers have learned that the expansion of censorship

---

15 Parkin, Siodhbhra, and Jiayun Feng. 2019. "China's #Metoo Movement, Explained." *SupChina*, July 12; Zhang, Han. 2018. "One Year of #Metoo: How the Movement Eludes Government Surveillance in China." *The New Yorker*, October 10.

16 Hernández, Javier C., and Iris Zhao. 2018. "Students Defiant as Chinese University Warns #Metoo Activist." *The New York Times*, April 24; Zhao, Wolfie. 2018. "#Metoo Movement Turns to Ethereum to Evade Censorship." *CoinDesk*, April 25.

17 The Ethereum transaction hash is 0x2d6a7b0f6adeff38423d4c62cd8b6c-cb708ddad85da5d3d06756ad4d8a04a6a2, sent on 23 April 2018.

can counterintuitively increase the amount of information available to citizenry. In September 2014, the photo-sharing website Instagram was suddenly blocked by the Great Firewall as part of the Chinese state's response to the 2014 Hong Kong protests. The blocking encouraged many of Instagram's large mainland Chinese userbase to download VPNs to continue using the service. While Instagram use tends to be apolitical, William R. Hobbs and Margaret E. Roberts show that in fact, the perverse consequence of this now expansive censorship avoidance was to increase mainland Chinese readership of Wikipedia pages on historical topics (such as the Tiananmen Square massacre).[18] Having been forced to liberate themselves from the Great Firewall, Instagram users started using that freedom to engage with information much more threatening to Chinese authoritarianism than the incidental photos of daily life that characterise most Instagram feeds.

Censoring the internet is costly. Censorship is relatively simple when there are only a few authorised newspapers, but extremely difficult in an age of social media self-publishing. There were around 2 million people employed as censors in government and private organisations in 2013, and all indications suggest that number has sharply increased since the spread of social media.[19] The similarly rapid rise of video content makes the task even more labour intensive. Unsurprisingly, the operation of the Great Firewall is increasingly being managed by AI.[20]

18 Hobbs, William R., and Margaret E. Roberts. 2018. "How Sudden Censorship Can Increase Access to Information." *American Political Science Review* 112 (3): pp. 621-636.

19 Cadell, Cate, and Pei Li. 2017. "Tea and Tiananmen: Inside China's New Censorship Machine." *Reuters*, September 29.

20 Yang, Yuan. 2018. "Artificial Intelligence Takes Jobs from Chinese Web Censors." *The Financial Times*, May 22.

Some content is relatively easy to identify automatically, like pornography (which is heavily censored in China). Political content is harder, as in response it tends to be obfuscated behind layers of suggestion and metaphor. One well-known example is how the children's book character Winnie the Pooh acts as a stand in for China's President Xi Jinping in critiques of Chinese authoritarianism; references to Winnie the Pooh are now censored, and the 2018 film *Christopher Robin* film was banned.[21] Navigating these complex cultural references is out of the skillset of current AI, and so censorship tends to be done by humans and automation together.

Strikingly, that interaction—between human and machine censors—presents an opportunity for liberty. Efforts to circumvent censorship in this context focus on analysing patterns about how the censorship regime functions and identifying the gaps through which information can be passed. Blockchain is one such obvious gap, at least to the extent that the Chinese state does not seek to prevent access to blockchain nodes (which, given the significance of blockchain as economic infrastructure, would be costly). Similar techniques exploit the fact that the Chinese economy is reliant on some Western services, like Amazon's cloud computing platform, AWS. GreatFire.org offers browsers and other applications to monitor censored content and provide unrestricted access to the internet for those behind the Great Firewall.[22] Its FreeBrowser application has been downloaded by more than 3 million

---

21 Haas, Benjamin. 2018. "China Bans Winnie the Pooh Film after Comparisons to President Xi." *The Guardian*, August 8.

22 GreatFire.org. 2018.

users, the vast majority of whom are Chinese.[23]

Other research has focused on using AI to determine what content is likely to be censored. The Great Firewall does not prevent all negative comments about the state from being shared on social media—rather, it seems to be focused on comments that might spur collective action.[24] Researchers have until recently focused on how satire can be used for censorship circumvention, how secret messages can be encoded within otherwise uncontroversial messages, and how using homophones (that is, words which sound the same but have completely different meanings) can avoid automated censors.[25] But these approaches tend to involve a lot of human work, making it both inaccessible to most users and complex to understand. In addition, research into the homophone approach to circumvention suggests that the technique adds a burden on censors of 15 additional hours per term.[26] But eventually culturally literate censors will identify and close-off these circumvention paths.

---

23 Banjo, Shelly, and Lulu Yilun Chen. 2019. "Digital Dissidents Are Fighting China's Censorship Machine." *Bloomberg Businessweek*, June 4.

24 King, Gary, Jennifer Pan, and Margaret E. Roberts. 2013. "How Censorship in China Allows Government Criticism but Silences Collective Expression." *American Political Science Review* 107 (2): pp. 1-18.

25 Safaka, Iris, Christina Fragouli, and Katerina Argyraki. 2016. "Matryoshka: Hiding Secret Communication in Plain Sight." *Paper presented at the 6th USENIX Workshop on Free and Open Communications on the Internet (FOCI), Austin, Texas, August 6*; Lee, Siu-yau. 2016. "Surviving Online Censorship in China: Three Satirical Tactics and Their Impact." *The China Quarterly* 228: pp. 1061-1080; Hiruncharoenvate, Chaya, Zhiyuan Lin, and Eric Gilbert. 2015. "Algorithmically Bypassing Censorship on Sina Weibo with Nondeterministic Homophone Substitutions." *Paper presented at the Ninth International AAAI Conference on Web and Social Media, Oxford, United Kingdom, May 26-29.*

26 Hiruncharoenvate, Chaya, Zhiyuan Lin, and Eric Gilbert. 2015 "Algorithmically Bypassing Censorship on Sina Weibo with Nondeterministic Homophone Substitutions."

So censorship circumvention is possible. But it is labour intensive, and new techniques have to be constantly developed to keep ahead of the censors. For censorship circumvention to be not simply possible, but widely available, we need to be able to automate circumvention. This is where developments in AI are exciting. AI systems can predict with increasing accuracy what information will be censored. For example, a team at Montclair State University has been using neural networks to identify common linguistic features of censored text on the Chinese microblogging website Sina Weibo, thus making predictions about whether any given text will be censored.[27] This is not simply an academic question. If censorship can be predicted, it can be evaded. This research project raises the possibility for the development of tools that make real-time, automated assessments of how information is currently being censored, how it can be modified to avoid that censorship, and feed the responses of the censors back into the analysis.

The Great Firewall of China describes a complex set of tools and technologies that apply the network layer (the internet infrastructure) up to the level of the application layer (such as social media services), used by both the Chinese state and private firms acting under obligation to

---

27 Ng, Kei Yin, Anna Feldman, and Chris Leberknight. 2018. "Detecting Censorable Content on Sina Weibo: A Pilot Study." *Paper presented at the Proceedings of the 10th Hellenic Conference on Artificial Intelligence, Patras, Greece, July 9-12*; Leberknight, Chris, and Anna Feldman. 2019. "Leveraging NLP and Social Network Analytic Techniques to Detect Censored Keywords: System Design and Experiments." *Paper presented at the Proceedings of the 52nd Hawaii International Conference on System Sciences, Hawaii, United States, January 8-11*; Ng, Kei Yin, Anna Feldman, Jing Peng and Chris Leberknight. 2019. "Neural Network Prediction of Censorable Language." *Paper presented at the Proceedings of the Third Workshop on Natural Language Processing and Computational Social Science, Minneapolis, Minnesota, June 6.*

the state. Tools like VPNs and Tor that work to circumvent the network layer are vulnerable when the government targets their bridges. In response, researchers are working on techniques to distribute and hide bridges and evade blocking.[28] Decentralised file-storage systems using blockchain (such as the InterPlanetary File System, or IPFS) offer alternative paths to access websites and software. When the Spanish courts forced the ISPs to block access to pro-Catalan referendum websites in 2017, the Catalan Pirate Party used IPFS to mirror the sites and provide uncensorable access.[29]

While it is theoretically possible for states to completely prevent their citizens from using these networks, doing so would potentially exclude them from exploiting key assets of the global economy. For example, one current oasis of free exchange of ideas allowed in China is the code sharing and version control management service GitHub. GitHub, which has been owned by Microsoft since 2018, is a core tool for modern software development, allowing developers access to massive open source libraries of code and private version control. Preventing access to GitHub would cripple the Chinese software industry. Periodic attempts have been made to do so, particularly targeting the GreatFire.org project.[30] But none of these have as yet been permanent. As a result, GitHub is a surprisingly fertile site for

---

28 Nasr, Milad, Sadegh Farhang, Amir Houmansadr and Jens Grossklags. 2019. "Enemy at the Gateways: Censorship-Resilient Proxy Distribution Using Game Theory." *Paper presented at the Network and Distributed Systems Security (NDSS) Symposium, San Diego, California, United States, 24-27 February.*

29 Balcell, Marta Poblet. 2017. "Inside Catalonia's Cypherpunk Referendum." *Eureka Street* 27 (20).

30 See, for instance: Kan, Michael. 2013. "GitHub unblocked in China after former Google head slams its censorship." *ComputerWorld*, January 23.

internal Chinese dissent, particularly around the punishing "996" labour practice (9:00 am to 9:00 pm, 6 days per week) that is common in the Chinese software industry.[31]

We have focused here on China because the Chinese regime has managed to exclude nearly a fifth of the world's population from the open internet, but internet censorship is globally widespread. The Freedom of the Net report by Freedom House identifies countries such as Iran, Ethiopia, Syria, Cuba, Vietnam, Uzbekistan, Pakistan, Saudi Arabia and Egypt with similarly repressive regimes. Freedom House estimates that of the 3.7 billion internet users globally, 55 per cent live in countries that prevent their users from viewing some political, social or religious content online.[32] The Chinese state runs workshops and seminars for foreign governments seeking to control public opinion and the news media, sometimes connected with the Belt and Road initiative.[33]

The use of AI for internet censorship is in its early stages. But so too is the use of AI for censorship circumvention. The challenge for censors is to allow as free access as possible to the global internet infrastructure—both because citizens demand it and because the digital economy is a key driver of economic growth—while at the same time preventing information harmful to the interests of the state. In most instances the Chinese state has preferred to block more rather than less—whole domains (i.e. Facebook.com) and multi-use keywords

---

31 Feng, Emily. 2019. "Github Has Become a Haven for China's Censored Internet Users." *NPR*, April 10.

32 Shahbaz, Adrian. 2018. "Freedom on the Net 2018: The Rise of Digital Authoritarianism." *Freedom House*.

33 Ibid.

from its local social media platforms (i.e. 'Winnie the Pooh'). When one Chinese stock index fell 64.89 points in 2012, censors responded by blocking words like "stock market" and "index" to prevent any political dissent. The number resembled too closely the date of Tiananmen Square massacre—June 4, 1989.[34] As the artist and dissident Ai Weiwei has written:

> China may seem quite successful in its controls, but it has only raised the water level. It's like building a dam: it thinks there is more water so it will build it higher. But every drop of water is still in there. It doesn't understand how to let the pressure out. It builds up a way to maintain control and push the problem to the next generation.[35]

The promise of AI is that it can probe for weaknesses in a censorship regime. Regardless of whether that regime is a human system or an AI-driven system (or a hybrid of both), paths for communication must be discovered and exploited. Such paths may well be temporary—and of course the censoring state has as much access to technological capability as those who would avoid censorship—but in this contest the censor is at a disadvantage. Information is non-excludable. It cannot be taken away once acquired.

It is too easy to overstate the effectiveness of China's Great Firewall already. For those who wish to surmount it is a costly and frustrating

34 Bradsher, Keith. 2012. "Market's Echo of Tiananmen Date Sets Off Censors." *The New York Times*, June 4.

35 Weiwei, Ai. 2013. *Weiwei-Isms.* Princeton, New Jersey: Princeton University Press.

burden, not a stark barrier. The Chinese state has made a critical design decision: that it wishes to raise the cost of its citizens accessing the public sphere, rather than prevent them from doing so at all. North Korea has chosen to close itself completely off from the open internet, and has done so at the expense of being able to exploit the vast scientific and economic benefits of global market capitalism. Margaret E. Roberts calls China's approach the "friction" model of information control.[36] But adding friction creates an uneven landscape which entrepreneurs can navigate to avoid and circumvent those controls. New, better tools, in the hands of a distributed global community that wish to open debate in closed societies, promise to lower the costs of censorship and guarantee freedom of speech even in those societies whose states oppose it.

CENSORSHIP RESISTANT ECONOMIES

The short messaging social media network Twitter was launched in the world in July 2006. Facebook was first opened to all users in September 2006. By 2016, these social media websites—particularly Facebook—were so widely adopted that there have been claims that the two most consequential democratic votes in that year (the 2016 presidential election in the United States and the British vote to leave the European Union) would not have happened as they did if it weren't for social media.[37] While the material impact that social media had on these elections is debatable, it is nonetheless true that the social media platforms have an extremely significant role in public debate. More

---

36 Roberts, *Censored: Distraction and Diversion inside China's Great Firewall.*

37 See, for example: Lapowsky, Issie. 2016. "Here's How Facebook Actually Won Trump the Presidency." *Wired*, November 15; Cadwalladr, Carole. 2019. "Facebook's Role in Brexit – and the Threat to Democracy." *TED Talk*, April.

than 1.6 billion people worldwide are active on Facebook daily—and 2.4 billion active at least once a month.[38] Twitter has 152 million daily active users.[39]

The centrality of these digital social media platforms has led commentators on both the left and the right to argue that they wield enormous power over politics and society. On the left, the Democratic presidential contender Senator Elizabeth Warren has called for Facebook, Google, and Amazon to be broken up with antitrust policy to restore competition.[40] On the right, a Republican senator, Josh Hawley, has similarly called for Facebook to be broken up and be divested of two of its services, the photo sharing site Instagram and the messaging application WhatsApp.[41]

For Hawley and many other conservatives, the charge sheet against social media platforms is longer again. In addition to the familiar arguments that these platforms harm privacy, one particularly sensitive criticism is that they are systematically biased against conservative voices, downgrading conservatives in search rankings, "deplatforming" them by removing conservatives from their platforms, or "demonetizing" them by removing their videos from advertising networks. As one conservative analyst wrote in the *Wall Street Journal*:

---

38 Facebook. 2019. "Facebook Q4 2019 Results." *Facebook.*

39 Specifically "monetizable" daily active users, a metric designed to exclude bots. Twitter. 2019. "Twitter Selected Company Metrics and Financials." *Twitter.*

40 Warren, Elizabeth. "Here's How We Can Break up Big Tech." *Medium,* March 8.

41 Gordon, Marcey. "Us Senator Asks Facebook Ceo Mark Zuckerberg to Sell Instagram and Whatsapp." *Business Insider Australia*, September 20.

> From the Ivy League to Hollywood and the mainstream media, Americans with traditional morals or conservative politics have often felt excluded from the country's elite, culture-creating institutions. Facebook offered an alternative: a place to express views and share news that you couldn't find in the *New York Times*.
>
> Over time, however, many conservatives lost trust in Facebook, believing it discriminated against them. The increasing scale and complexity of Facebook's content-moderation practices made matters worse. In 2016 Facebook employees were accused of suppressing conservative articles in the news feed's now-discontinued "trending" section. In April 2018 Congress grilled Facebook chief executive Mark Zuckerberg about anticonservative bias, from blocked content to suspended accounts.[42]

In 2019 the White House even launched a dedicated webpage for social media users to let the government know if they had been "suspended, banned, or fraudulently reported" from these platforms as a result of "suspected political bias."[43] Nonetheless, this is hardly a concern of the right. The Electronic Frontier Foundation's Jillian York has argued that:

> The companies that host the vast majority of our online expression

42 Kyl, Jon. 2019. "Why Conservatives Don't Trust Facebook." *The Wall Street Journal*, August 20.

43 The White House. "White House Type Form." See also: Heater, Brian. 2019. "The White House Wants to Know If You've Been 'Censored or Silenced' by Social Media." *TechCrunch*, May 16.

> – most only a little over a decade old – have amassed a tremendous amount of power, and the ability to influence everything from how we consume news to what we wear … it is also vital that we acknowledge the harm that handing over governance to corporations – whose primary interest is profit – can engender.
>
> But with power already in their hands, it's nevertheless important that we push companies toward meaningful transparency and accountability. For starters, this means opening up their processes for public scrutiny; offering due process to users whose content is taken down; and being straight with users about how their data is collected and used.
>
> As we look toward solutions, we must prioritize democratic norms above profit maximization.[44]

It's not obvious that individual accounts being banned from social media platforms should be described as "censorship," or even an attack on freedom of speech. The right to speak is not the right to the same as to speak on someone else's private platform. The traditional liberal view has always been that freedom of expression is a right asserted against censorship by the government, or against the use of the legal system to suppress or punish speech.[45] Historically, many left wing and socialist commentators claimed in the nineteenth and twentieth century there was no functional freedom of speech because a small number of

---

44 York, Jillian. 2018. "Facebook: The New Censor's Office." *New Internationalist*, April 17.

45 See, for instance, this exploration: Gurri, Adam. 2019. "No One Is Owed an Audience: Facebook, Free Speech, and Liberal Tolerance." *LiberalCurrents*, October 29.

(typically conservative) media moguls controlled the press.[46]

In response, liberals argued that freedom would be secured by ensuring that there was no legal restriction on anyone establishing their own media organisation, and that to force a private company to host speech that it did not agree with would itself be a violation of the speech rights of those who owned that company.[47] The claim that modern social media networks are threatening free speech by preventing some voices from using their platforms is a conservative appropriation of an old progressive argument—that a right to free speech is meaningless without access to the corporate-run media, and therefore the state should intervene.

Is the old progressive argument any more valid in the age of social media? As Adam Thierer pointed out in his 2005 book *Media Myths: Making Sense of the Debate over Media Ownership,* these claims were extremely sensitive to technological constraints. Over the last two decades we have seen an explosion of online content, available everywhere. The social media networks only form a small fraction of that content. In a time where the pro-communism website *Jacobin* and the right-populist website *Breitbart* are equally accessible, the idea that large firms are suppressing content diversity is obviously unsustainable.

---

46 The classic statement of this is: Herman, Edward S. and Noam Chomsky. 2010. *Manufacturing Consent: The Political Economy of the Mass Media.* Penguin Random House. See also: Johnston, Carla B. 2016. *Screened Out: How the Media Control Us and What We Can Do About It: How the Media Control Us and What We Can Do About It.* Routledge; Bagdikian, Ben H. 2004. *The New Media Monopoly.* Penguin Random House; McChesney, Robert Waterman, Robert W. McChesney, and John Nichols. 2002. *Our Media, Not Theirs: The Democratic Struggle against Corporate Media.* New York: Seven Stories Press.

47 Thierer, Adam. 2005. *Media Myths: Making Sense of the Debate over Media Ownership.* Washington, DC: The Progress & Freedom Foundation.

Where it was expensive in the print era to establish a niche magazine or newspaper, the cost of setting up a niche website is trivially small.

But even so, much of the public discourse has moved away from the independent blogs and websites of the 2000s and onto platforms—businesses that run multi-sided markets, providing users free or low cost access to their platform who they match to targeted advertising. And while these platforms are open—insofar as they are not curated, edited and controlled as the legacy media were—they still have the ability to remove content and ban users as they see fit. And it is in the use of that capability where the controversy has arisen.

Yet while they are distinct from the print mastheads of the past, this new generation of platforms are just as vulnerable to technological change. Digital platforms have the power to remove content and ban users because of the way they are governed. Facebook is fundamentally a corporation, with a governance structure that centralises and concentrates decision-making—ultimately up to its chief executive, Mark Zuckerberg. While Zuckerberg is partially constrained by his need to maximise shareholder value, he has a high level of discretion in how to do so. If that involves removing specific individuals or forms of content from the social network he oversees, then he has the ability—even a mandate—to do so. Those decisions could be politically motivated, or motivated by moral concerns.

This is where new technological approaches provide an opportunity. Entrepreneurs and developers have been experimenting with and building new generations of *decentralised* platforms—digital protocols that exploit blockchain technology to build distributed social media networks without any central authority. One prominent example is Steemit, a decentralised social media platform built on the dedicated Steem blockchain. Each individual post—as well as other interactions like comments and votes on the inbuilt voting system—is recorded on the blockchain. A native cryptocurrency token is designed to allow popular posts to be rewarded.

A blockchain here provides two critical features. First, these decentralised platforms are resistant to external censorship—their immutability prevents outside authorities from forcing content to be deleted or preventing users accessing and interacting on the platform. Second, decentralised platforms have distributed governance, which means there is no single agent that can make decisions about what is or is not allowed on the platform. Decisions about how to alter the network must be made by consensus—consensus between nodes that might be hosted across different legal jurisdictions and geographic locations.

The significance of this latter feature became evident in early 2019 when a hacker group calling itself TheDarkOverlord claimed to hold sensitive documents concerning the September 11 terrorist attacks and demanding ransom from "airlines, government agencies, the dozens of solicitor firms, the insurers, and the many others."[48] Many users of Steemit were surprised when user TheDarkOverlord was banned for violating the Steemit terms of service. However, the group's posts remained on the underlying Steem blockchain. Steemit.com is an accessible website for accessing the Steem blockchain run by a real company, Steemit Inc., with real employees, based in New York. The company runs legal risk for information posted on the website it controls. But the company has no control over what is placed on the blockchain itself, and anyone interested in looking at

48 O'Flaherty, Kate. 2019. "Who Is the Dark Overlord Threatening to Leak Sensitive 9/11 Documents?" *Forbes*, January 2. For the content see TheDarkOverlord. "Press Release 02 – Crypto-Cash for Crypto-Cache."

what TheDarkOverlord has posted is able to access the blockchain directly, if not particularly conveniently.

Indeed, this distributed governance mitigates another problem with the centralisation of management discretion on digital platforms. Claims by politicians and regulators that digital platforms hold excessive power in the market leave those platforms vulnerable to political interference and regulation. The Congress of the United States can drag Mark Zuckerberg in to answer questions from lawmakers, demanding that Facebook's policies around privacy, corporate consolidation, or content censorship are changed. Senator Josh Hawley was able to deliver his message that Facebook should divest Instagram and WhatsApp to Zuckerberg in a meeting, in person.[49] This is not possible when platforms have been decentralised—when there is no high profile individual, or just any particular individual, with the ability to make strategic decisions on the platform's behalf.[50]

Decentralising technologies change many of the fundamental assumptions behind our regulatory and legal system, not just freedom of speech. Consider, for instance, the role of antitrust in an economy where decentralised protocols dominate. Competition regulators around the world argue that Facebook and Google dominate certain advertising markets.[51] If decentralised protocols replace these centrally governed platforms (and, hypothetically, gained similarly dominant

---

49 Gordon, "Us Senator Asks Facebook Ceo Mark Zuckerberg to Sell Instagram and Whatsapp."

50 On competition policy and decentralised platforms, see recently Schrepel, "The Theory of Granularity: A Path for Antitrust in Blockchain Ecosystems."

51 The basis for this is sometimes questionable, see: Allen et al., "Submission on the Final Report of the Australian Competition and Consumer Commission's Digital Platforms Inquiry."

market share) competition regulators would have little recourse. With no Mark Zuckerberg to pull in front of a legislative committee, and no incorporated entity to enforce divestment or other regulatory decisions upon, competition law becomes defanged. Blockchain's censorship resistance is also regulatory resistance. Regulators may be able to target businesses that use the blockchain, or the interfaces between the real world and the blockchain world, but have no power to directly affect economic activity on chain.

One of the consequences of the digitisation of the economy is the expansion of the power of speech. Speech is no longer restricted to expressing ideas and convincing others. Through computer code—literally strings of digital characters—speech can act. Code can be money, as we have seen in cryptocurrencies. Code can be business models, as we have seen with distributed autonomous organisations. Code can be self-enforcing legal agreements, as we can see in smart contracts. If we can distribute that speech on censorship resistant networks, it is not only freedom of speech that is empowered, but the freedom to act too.

06.

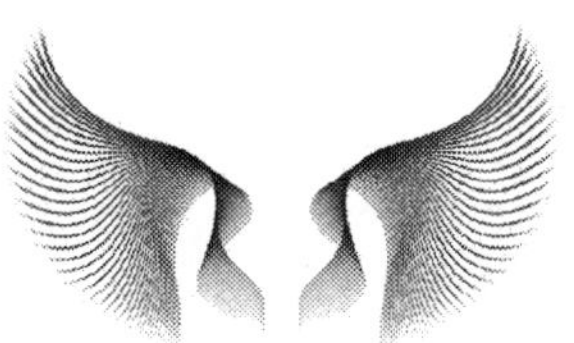

THE FUTURE OF PRIVACY

It would be hard to find an individual liberty that better epitomises the tension between private autonomy and state power than privacy. A key part of the classical liberal vision of the world is the fundamental difference between public life (the space of society, of politics, of communities and nations) and private life (the space of pleasure, intimacy and retreat).[1] As Hayek wrote in the *Constitution of Liberty*,

> recognition of a protected individual sphere has in times of freedom normally included a right to privacy and secrecy, the conception that a man's house is his castle and that nobody has

1 Berg, *The Classical Liberal Case for Privacy in a World of Surveillance and Technological Change.*

a right even to take cognizance of his activities within it.[2]

Repressive states work hard to erode this distinction. They are threatened by the very idea that there is a domain that government may not intrude upon. The worst dictatorships of the twentieth century were dictatorships of surveillance. In Soviet-ruled Russia, families were herded into shared apartments where neighbours watched each other, and reported misdemeanours to the authorities.[3] The Khmer Rouge, one of the most brutal regimes in human history, moved populations into communal housing, where they ate communal meals and were married in communal ceremonies. One Khmer Rouge slogan—command—was "Do not harbour private thoughts!"[4]

But if privacy is a fundamental liberty, it seems to be a liberty in retreat. Liberal democratic states deploy surveillance devices in unparalleled quantities. One estimate suggests that there are more than 620,000 closed circuit television (CCTV) cameras in London, meaning that there is one for every fourteen citizens.[5] The drive towards "smart" cities—city planning informed by data analytics—has led to the distribution of vast quantities of sensors that measure things like traffic and pedestrian flows and how buildings and open spaces

---

2 Hayek, Friedrich A. 1960. *The Constitution of Liberty: The Definitive Edition.* Chicago, Illinois: University of Chicago Press. p. 209.

3 Figes, Orlando. 2008. *The Whisperers: Private Life in Stalin's Russia.* London, UK: Penguin Books; Attwood, Lynne. 2017. *Gender and Housing in Soviet Russia: Private Life in a Public Space.* Manchester, UK: Manchester University Press.

4 Locard, H. 2004. *Pol Pot's Little Red Book: The Sayings of Angkar.* Chiang Mai, Thailand: Silkworm Books.

5 Ratcliffe, Jonathan. 2019. "How Many CCTV Cameras Are There in London 2019?" *CCTV.co.uk*, May 29.

are used. These cameras are increasingly "smart"—traffic cameras can now often identify license plates (and match them against other databases), and a new generation of CCTV cameras have facial recognition capabilities.

Only a fraction of these surveillance devices are owned or controlled by governments. Indeed, the vast majority are deployed by the private sector. Private sector CCTV cameras in London may outnumber government CCTV by as much as 70 to 1.[6] Some of these are required by law (for instance, some cities require nightclubs to have cameras at their entries and exits) but most operate to prevent property crime.

The picture of privacy rights as we enter the third decade of the twenty-first century is one in which both government and corporate actors are encroaching on our private domain. States survey their populations to monitor everything from national security threats to misdemeanour driving offenses. At the same time, however, we disclose vast amounts of information to the firms that we interact with on a day to day basis. Not only do we post about our lives on social media, but our mobile phones broadcast our location to mapping providers, telecommunications providers, they store intimate data about our relationships and preferences, record the websites we have visited, and the video clips we may have watched.

The two are, of course, not at all morally comparable. The coercive state surveillance revealed in 2013 by the whistleblower Edward Snowden was unlawful, involuntary, and expansive. The voluntary sharing of personal information with technology companies is done

6 British Security Industry Association. 2013. "The Picture Is Not Clear: How Many CCTV Surveillance Cameras Are There in the UK?" *British Security Industry Association.*

as part of the consumption of the product, or in return for a service. The classical liberal approach to privacy recognises a clear difference between these two situations and organisational forms. Nonetheless, it is the case that we simultaneously worry about the state observing our internet activity and, at the same time, happily install smart home devices like Google Home or Alexa that passively listen to our intimate conservations at home.[7]

No wonder then that it is common to hear that privacy is dead—or, at least, terminally ill. Modern states and firms rely on access to vast quantities of data in order to deliver services, market products, assess credit, and customise advertising. While that data is usually (and sometimes by law) acquired solely for the purposes of delivering business services, it can be reconstituted by malicious actors to offer a deeply personal window into someone's life. As one scholar has written, even just an internet search history can offer a "metaphorical X-ray photo of one's thoughts, beliefs, fears, and hopes."[8] Given that the data constituting those search histories are used by our browser and our internet service providers, they travel through the global internet infrastructure, and in most developed countries are stored for access by regulators and law enforcement (sometimes accessible without a warrant), it is easy to be pessimistic.

Accidental and malicious disclosures of personal data can be highly damaging. The data breach that affected the dating service Ashley Madison in 2015 was particularly destructive because the website

---

7 Sometimes not so passively – see for instance: Day, Matt, Giles Turner, and Natalia Drozdiak. 2019. "Amazon Workers Are Listening to What You Tell Alexa." *Bloomberg*, April 11.

8 Tene, Omer. 2008. "What Google Knows: Privacy and Internet Search Engines." *Utah Law Review*: pp. 1433-54.

catered entirely to people who were looking for extramarital affairs. Databases that are stored for the purposes of law enforcement or the health system are regularly the subject of unauthorised access. In one particular case in 2019 in the Australian state of Queensland, a police officer unlawfully looked up the address of a domestic violence victim and then passed that information to her former partner.

But the litany of privacy violations that fill out our press are misleading about the future of privacy. A new generation of technologies promise to radically reshape the landscape between private information and the public domain, with potentially dramatic consequences not only for how we protect our personal information, but for how government itself functions. Many of the building blocks of that revolution are already available to consumers—they can be used on our computers and phones with just a few keystrokes. But to understand these new technologies of privacy, we have to look at how the world has evolved since the revelation of one of the most significant scandals in the history of privacy—when the whistleblower, Edward Snowden, revealed in 2013 the vast apparatus set up by the National Security Agency to monitor the internet communications of the entire planet.

The September 11 terrorist attacks in 2001 revealed serious weaknesses in the intelligence apparatus of the United States, in part because the intelligence community was still organised around the threat of superpower competition inherited from the Cold War.[9] In the wake of the attacks, the National Security Agency (NSA), the country's signals intelligence agency, made a concerted effort to expand its monitoring of telephony and internet traffic. Between 2005 and 2014

---

9 Zegart, Amy B. 2006. "An Empirical Analysis of Failed Intelligence Reforms before September 11." *Political Science Quarterly* 121 (1): pp. 33-60.

the NSA was led by General Keith B. Alexander, who oriented the agency around massive information collection. As one of his former colleagues told *Foreign Policy*: "Alexander's strategy is the same as Google's: I need to get all of the data."[10]

Enabled by an extremely broad interpretation of the 2001 Patriot Act the NSA, under Alexander's leadership, required American telephone providers to hand over complete call data records to the NSA so it could be searched and interrogated—regardless of whether the calls related to terrorist suspects or threats.[11] Similar arrangements were made with internet service providers, digital platforms, software companies, telecommunications infrastructure providers, and hardware manufacturers. The result being that the NSA was collecting massive and comprehensive records about global communications that the agency could then access at their leisure in the course of investigations.

The size and expansiveness of this program was revealed in 2013 when Edward Snowden, a computer systems manager, who had variously been employed directly by the CIA and as a private sector contractor to the NSA, leaked internal NSA documents to a number of prominent outlets including *The Guardian, Der Spiegel*, and *The Washington Post*. The leaks sparked a global outrage, and a large-scale debate about internet security in an environment where liberal and democratic governments presented a threat to privacy. Nonetheless

---

10 Harris, Shane. 2013. "The Cowboy of the NSA." *Foreign Policy*, September 9.

11 A useful overview of the political and legal basis of the program is provided by: Sanchez, Julian. 2014. "Snowden: Year One." *Cato Unbound*, June 5. See also: Greenwald, Glenn. 2014. *No Place to Hide: Edward Snowden, the Nsa, and the U.S. Surveillance State.* Henry Holt and Company; Edgar, Timothy H. 2017. *Beyond Snowden: Privacy, Mass Surveillance, and the Struggle to Reform the NSA*. Brookings Institution Press; Snowden, Edward. 2019. *Permanent Record.* UK: Pan Macmillan.

policy changes in response to that outrage was minimal. The legal framework that underpinned the NSA's bulk data collection remains in place, and several governments, including Australia, have since introduced further legal frameworks requiring bulk data retention by telecommunications providers for the purposes of law enforcement.[12]

Yet the strongest response to the Snowden revelations came from the private sector. Snowden had dramatically revealed a close cooperative relationship between technology companies and the national security state. Public pressure—or even just public awareness—has severed that relationship, at least for many of the most prominent, consumer-facing firms. As the cryptographer Matthew Green writes, "It's also easy to forget how much things have changed" since the Snowden revelations.[13] Before 2013, vast swathes of internet and telecommunications messages were sent in plaintext—that is, without the benefit of being encrypted so that only those who held a key to the message could access it. While Snowden revealed that the NSA had the capability to break the encryption of many common services, a critical share of digital communication wasn't even using that.

Thus the years after the Snowden revelations saw the launch of a large number of new services that dramatically expanded user privacy over internet communication. For example, digital platforms like Google encouraged website administrators to convert from the HTTP protocol (the basic communications standard used by web pages) to a secure form of the protocol (HTTPS), where information shared between the

---

12 Suzor, Nicolas P., Kylie M. Pappalardo, and Natalie McIntosh. 2017. "The Passage of Australia's Data Retention Regime: National Security, Human Rights, and Media Scrutiny." *Internet Policy Review* 6 (1).

13 Green, Matthew. 2019. "Looking Back at the Snowden Revelations." *A Few Thoughts on Cryptographic Engineering*, September 24.

browser and the website is encrypted. Another major change was when the text messaging service built into mobile telephone communications (that is, SMS, or short messaging service) was, over time, supplanted by messaging systems that offered end-to-end encryption—where access to the content of communications is only available to sender and recipient (not messaging service providers). In 2020 there is now a wide and growing range of centralised and decentralised messaging applications—for instance Signal and Keybase—that are beginning to satisfy the demand for secure communications.

Since the birth of computing, person-to-person communications has been riddled with security weaknesses. Email encryption, for instance, has never caught on at scale.[14] So the sudden and rapid post-Snowden adoption of encrypted communication represented a fundamental shift in the underlying architecture of the internet. No lesser authority than the US government recognised the connection between the NSA leaks and rapidly changing security practices. In 2016 the US Director of National Intelligence, James Clapper, claimed that "As a result of the Snowden revelations, the onset of commercial encryption has accelerated by seven years."[15] But possibly more fundamental, as Matthew Green writes, was a change in attitude across both industry and consumers:

---

14 LaFleur, Kendal Stephens, and Lei Chen. 2014. "Email Encryption: Discovering Reasons Behind Its Lack of Acceptance." *Paper presented at the Proceedings of the International Conference on Security and Management (SAM), Las Vegas, July 21-24.*

15 McLaughlin, Jenna. 2016. "Spy Chief Complains That Edward Snowden Sped up Spread of Encryption by 7 Years." *The Intercept*, April 26.

> The idea that governments would conduct large-scale interception of our communications traffic was a point of view that relatively few "normal people" spent time thinking about — it was mostly confined to security mailing lists and X-Files scripts. Sure, everyone understood that government surveillance was a thing, in the abstract. But actually talking about this was bound to make you look a little silly, even in paranoid circles.
>
> That these concerns have been granted respectability is one of the most important things Snowden did for us.[16]

Indeed, it is the post-Snowden context of rapid adoption of strong encryption that has led to the new panic from governments in the developed world that law enforcement agencies are now "going dark"—that is, are no longer able to access privately held information, particularly that held on individual devices. This concern has been of such significance that some governments have sought to undermine strong encryption through the imposition of "backdoors" (paths to accessing encrypted information that only law enforcement agencies can utilise) or, as in Australia, the introduction of legislation that requires technology companies to modify their software according to law enforcement agency whims.[17]

The harm of these policies is not solely to the right to privacy. Encryption protects users of digital technology against not only the overweening state, but other states, hostile actors, and digital criminals. It is not possible to undermine encryption in such a precise way that

---

16 Green Looking Back at the Snowden Revelations.

17 On the Australian example see: Newman, Lily Hay. 2018. "Australia's Encryption-Busting Law Could Impact Global Privacy." December 7.

it only allows the "good guys" in while still keeping the "bad guys" out. It is one thing for advocates of encryption backdoors to claim that we ought to trust "our" governments (in the West, our liberal and democratic governments) to only use their power for "good." But do we also trust the governments of the unfree world—who, after all, have as much access to the global internet that our computers are connected to? Digital device security is a major and increasing technical and social problem. Undermining the efforts of technology providers to protect their users makes us all less safe.[18]

This dispute between encryption advocates and technology firms on one side and law enforcement and national security agencies on the other has more significance than first may appear. Contrary to the hyperbolic claims of law enforcement, we in fact live in what the academic Peter Swire describes as the "golden age" of surveillance.[19] As our devices and internet activities record more and more information about us, those records become a target for both state and non-state eavesdroppers. At no time before in history have law enforcement agencies been able—as they now can—to demand from telecommunications and technology companies, for example, a complete record of our travel, triangulated by our always-on GPS-enabled mapping software and constant pinging of cell towers. At no time before in history have such a complete map of our every thought been recorded through our internet searches, emails, and text messages, our interactions with smart home devices, the CCTV cameras that are scattered

---

18 Abelson, Hal et al. 1997. "The Risks of Key Recovery, Key Escrow, and Trusted Third-Party Encryption." *World Wide Web Journal* 2 (3): pp. 241-257.

19 Swire, Peter. 2015. "Going Dark: Encryption, Technology, and the Balance between Public Safety and Privacy." *Hearing by Senate Judiciary Committee*, July 8.

throughout our cities, our ride-sharing apps … we could go on. It is trivially easy for state agents to access these records. Australia's data retention scheme, for example, which requires internet service providers to store metadata on internet activity for at least two years, does not require law enforcement officials to get a warrant to access some of the most personal data—who we have been emailing, the websites we communicate with, and so on.

In that context, tools that prevent unauthorised access to information—information that we as consumers and citizens want to protect from the state, from corporations, from employers, colleagues, even friends and loved ones—even in the uneven and uncertain ways these tools have been implemented, offer a significant weapon in the arsenal of liberty. When the classic ideas about privacy were being first developed in the first half of the twentieth century, the question was what legal framework would best protect this complex right.[20] As we approach the third decade of the twenty-first century, the question is how best to use technologies to take our privacy back—and how the state should respond.

PRIVACY WITHOUT SECRECY

In 1985 the computer scientist David Chaum published a paper in the journal *IEEE Security & Privacy* titled, "Security without Identification: Transaction Systems to Make Big Brother Obsolete."[21] In the mid-1980s, widespread internet use was at least a decade away. But as Chaum presciently wrote, "Computerization is robbing individuals of the ability to monitor and control the ways information about them

20 For example: Westin, Alan F. 1967. *Privacy and Freedom.* New York: Atheneum.

21 Chaum, David. 1985. "Security without Identification: Transaction Systems to Make Big Brother Obsolete." *Communications of the ACM* 28 (10).

is used."[22] Chaum foresaw two possible futures. One in which the digitization of the economy led to economic centralisation—large firms and governments gathering massive banks of information about individuals, and using that to grow in size and influence. But the other future was one in which personal information could be "partitioned into separate unlinkable relationships" using cryptography to obscure the relationship between an individual and their personal data. Doing so might not only reverse the centralisation that has been driven by computerisation, but could be radically democratising.

Indeed, Chaum's essay was one of early documents of what came to be known as the "cypherpunk" movement—a play on the cyberpunk term, which emphasised the political *and* economic role of cryptography. These early cypherpunks argued that protecting information behind cryptography represented a radical shift in the relationship between individuals and organisations. As Chaum concluded his essay:

> Advances in information technology have always been accompanied by major changes in society: The transition from tribal to larger hierarchical forms, for example, was accompanied by written language, and printing technology helped to foster the emergence of large-scale democracies. Coupling computers to telecommunications technologies creates what has been called the ultimate medium—it certainly is a big step up from paper. One might ask, To what forms of society could this new technology lead? The two approaches appear to hold quite different answers.[23]

Other cypherpunks took this claim further. In his 1988 "Crypto

22 Ibid., p. 1030.

23 Ibid., p. 1044.

Anarchist Manifesto," Timothy May compared the spread of cryptography in digital communications to the evolution of property rights:

> Just as the technology of printing altered and reduced the power of medieval guilds and the social power structure, so too will cryptologic methods fundamentally alter the nature of corporations and of government interference in economic transactions. Combined with emerging information markets, crypto anarchy will create a liquid market for any and all material which can be put into words and pictures. And just as a seemingly minor invention like barbed wire made possible the fencing-off of vast ranches and farms, thus altering forever the concepts of land and property rights in the frontier West, so too will the seemingly minor discovery out of an arcane branch of mathematics come to be the wire clippers which dismantle the barbed wire around intellectual property.[24]

In its early years, Bitcoin, which came firmly out of this cypherpunk milieu, seemed to be the manifestation of these ideological goals. Bitcoin offered a digital currency that was independent of both the state with its central bank and the private corporate banking system. Users interacted only through digital addresses—peer to peer, with no central authority in charge. This was obviously an institutional shock. When government authorities first encountered cryptocurrencies their major concerns were whether cryptocurrencies represented the end of financial oversight and tax surveillance. It is still common to hear government officials ask whether financial transactions made in Bitcoin

---

24 May, Timothy C. 1992. "The Crypto Anarchist Manifesto." November 22.

can be taxed successfully.

But for all the innovation that Bitcoin represents, it does not represent a watershed in the privacy of financial transactions. Indeed, as we pointed out in Chapter 2, Bitcoin's model of trust comes from its public nature—any user of the Bitcoin network can, if they choose, audit each and every transaction on its ledger, stretching back to the first Bitcoin transaction by Satoshi Nakamoto. While Bitcoin addresses are not natively linked to the identity of people in the real world, in the decade since the network was launched tax and law enforcement agencies have found it relatively simple to trace transactions and attribute addresses to individuals. Indeed, Bitcoin transactions are significantly less anonymous than traditional physical currency—that is, cash.

So unsurprisingly as entrepreneurs and hobbyists started playing with the structure of cryptocurrencies in the shadow of Satoshi, privacy was one of the first features they sought to build into the design. Early approaches involved the use of Bitcoin "mixers," or "tumblers," which obfuscate where transactions originate and where they are sent by mixing them up with other transactions. The digital currency Monero, launched in 2014, embeds privacy into the protocol itself: the sender, the recipient and the amount of the transaction are private, even while Monero remains a publicly auditable ledger.

Another privacy-focused cryptocurrency, Zcash, uses a type of zero-knowledge proof in order to hide the details of transactions from third parties while remaining a public and open blockchain. Zero-knowledge proofs are a type of mathematical proof that allows one participant in an exchange to prove to another participant that they know a certain fact, without needing to reveal the fact itself. It is a way to prove knowledge without revealing the content of that knowledge.

Zero-knowledge proofs offer a radical re-envisioning of the

protection of privacy. There are many situations in life where we want to prove something. For example, say that we have a right to drive, or are over the drinking age—where, by necessity, we have to share a drivers licence or other ID to do so. But when we hand over our identification card, we are also handing over much more than just whether we have the right to drive—we are sharing with a police officer or club bouncer our names, our addresses, and our precise date of birth. This is much more information than is necessary to prove the bare facts of our right to drive or drink. Zero-knowledge proof systems allow us to drastically reduce the amount of information we share.

What these technologies offer is a vision of a society where we have to exchange far less personal information while continuing to fully engage in that society. This is a potentially radical change. Privacy choices have always been trade-offs between revealing information about ourselves and participating in the wider world. We must share to exchange. If we wish to enter a club, we have to share our identification document. If we want credit from a bank, we have to share our financial history. But privacy-enhancing technologies allow us to exchange at a vastly lower cost (in sharing information).

Privacy technologies allow us to rethink how we engage with firms, governments, and civil society organisations. At present we have to reveal information to organisations, but we might not want specific individuals to access that information. Banks want to know whether we are creditworthy, but we don't necessarily want an individual person to be able to scrutinise our individual transactions. Health insurers want to know whether we have had any covered medical procedures, but we don't want to share private details with an individual employee of that insurer. The government wants to know whether we are in a certain tax bracket—but we don't want to share all the details of our income with an individual tax employee. The government wishes to know if we are eligible for a disability payment—but we may not wish to share the specifics of our disability. Our current approach to economic exchange requires us to share a lot of information that is redundant to the transaction. These new technologies allow us to drastically reduce the amount of information we share while simultaneously engaging with the world around us.

That change is significant. Too often the choice for those who wish to protect their privacy has been to do so by withdrawing from social and economic interaction. To refuse to share information has been to decline to engage. A consumer who objects to the collection of their financial data by a bank will not be able to borrow from that bank. To keep information private it has, until now, needed to be kept secret. But these new technologies allow consumers to separate the two. Privacy can be no longer synonymous with secrecy. To protect information is not the same as hiding that information.

Privacy is desirable; so too is engagement with the world. When we make an exchange, our partners insist that we reveal information about ourselves—our credit worthiness, or our medical history—because there is a risk we might pretend to be something we are not. Privacy is hard to maintain in an environment where people do not fully trust each other. The new technologies of privacy offer us the possibility that entrepreneurs and developers can create privacy with trust.

PRIVACY AS A TASK FOR INNOVATORS

It is easy to be pessimistic about privacy in the digital age. Our digital devices create and store vast quantities of information about what we do in the world. An outside observer with unfettered access to the data created by the applications and services that are in essentially constant use—mapping and GPS, web browsers, email and calendar, chat and messaging services, social media, smart lights and heaters, streaming music and video, smart watches that measure heartbeat and sleeping patterns, cloud storage services, and on and on—could reconstruct not only our activities, but our moods, obsessions, even take a strong guess at what we have been thinking about from minute to minute. Then there are the devices owned by others—the CCTV cameras, the traffic management devices, the digital public transport tickets. This extraordinary array of services has suggested to many commentators that privacy is dead; that our only option is to adjust to a world without it, rather than complain pointlessly about a liberty that cannot be revived.[25]

Regulatory attempts to introduce privacy have been for the most part futile. Ambitious policies like the European Union's General Data Protection Regulation, an attempt to construct a legal right to personal data protection with quasi-global reach, have had a host of unintended consequences and problems. These have included extremely high compliance costs, corporate consolidation (as large firms are able to bear the costs of compliance more than small firms) and reduced innovation.[26] Researchers have determined that in some circumstances, for those who do not use the privacy protecting rights of the GDPR, privacy

---

25 See, for example: Morgan, Jacob. 2014. "Privacy Is Completely and Utterly Dead, and We Killed It." *Forbes*, August 19; Mance, Henry. 2019. "Is Privacy Dead?" *Financial Times*, July 19.

26 See: Thierer, Adam. 2018. "How Well-Intentioned Privacy Regulation Could Boost Market Power of Facebook & Google." *The Technology Liberation Front*, April 25; Thierer, Adam. 2018. "GDPR Compliance: The Price of Privacy Protections." *The Technology Liberation Front*, July 9; Allen, Darcy W.E. et al. 2019. "Some Economic Consequences of the GDPR." *Economics Bulletin* 39 (2): pp. 785-797.

outcomes have worsened.[27] And of course, government regulatory approaches to privacy do nothing to protect us against surveillance by the government itself.

But as we've explored in this chapter, a range of new technologies show that this pessimism about our privacy now and in the future is unwarranted. The history of privacy also provides a guide and reason not just for optimism, but to inspire innovation. Innovation in social and communications technologies tend to follow a clear path. At first they create new privacy risks.[28] But as they are adopted for more and more critical uses, innovators and entrepreneurs build privacy preservation onto them.

The history of the telegraph provides the prototypical example of this privacy-second dynamic. An electric telegraph pulses electricity through a wire, allowing communications by encoding information using the pulses and the silence between them. The standardised encoding came to be Morse code, which converts human language into a series of dots and dashes. The economic importance of the spread of telegraphy is hard to overstate. Where the spread of the railroad jumpstarted a national trade in goods, the telegraph represents the first buds of the information economy.[29] Not for nothing has the author

---

27 Aridor, Guy et al. 2020. "The Economic Consequences of Data Privacy Regulation: Empirical Evidence from GDPR." *SSRN 3522845*.

28 For an extended discussion of this dynamic, see: Berg, *The Classical Liberal Case for Privacy in a World of Surveillance and Technological Change*.

29 DuBoff, Richard B. 1980. "Business Demand and the Development of the Telegraph in the United States, 1844–1860." *Business History Review* 54 (4): pp. 459-479.

Tom Standage described the telegraph as the Victorian internet.[30]

But for all its possibilities, the telegraph was incredibly insecure. Intercepting messages was as simple as listening into the wire. This could be done with non-specialist equipment (one pre-Civil War story features the use of an iron poker to send messages and a wet tongue to receive them).[31] The business need for telegraph offices and their employees to manage the communication meant that surveillance was as easy as bribing a corruptible telegraphist. And the telegraphists themselves had unfettered access to communications, as a small subgenre of short stories and novellas, such as Henry James' 1898 *In the Cage*, emphasised.[32] These weaknesses meant that the economic potential of the telegraph was limited—firms that wished to pass on confidential information could not be certain that information would not leak to their competitors.

It was this problem that spurred the development of cryptography. As David Kahn writes, the "telegraph made cryptography what it is today."[33] Of course, the idea of encoding messages to prevent anybody but the intended recipient from reading them was not new: substitutional ciphers, where letters are switched for other letters, dates back well into the ancient world, and the most famous of these is known as the Caesar cipher. But with the telegraph came a sudden, concerted

---

30 Standage, Tom. 1998. *The Victorian Internet.* New York, US: Bloomsbury Publishing.

31 Petersen, Julie K. 2007. *Understanding Surveillance Technologies: Spy Devices, Their Origins & Applications.* New York: Auerbach Publications.

32 James, Henry. 1898. *In the Cage.* London, UK: Duckworth and Co.

33 Kahn, David. 1996. *The Codebreakers: The Comprehensive History of Secret Communication from Ancient Times to the Internet.* New York: Scribner's and Sons. At p. 189.

effort to innovate on the techniques for private communications. As Kahn writes, the 1800s saw a cottage industry of professional and amateur cipher makers:

> The great and widely felt need for secrecy awakened the latent interest in ciphers that so many people seem to have, and kindled a new interest in many others. Dozens of persons tried to dream up their own unbreakable ciphers … A surprising number of these dabblers were intellectual and political leaders of the day who beamed their powerful and original minds on the engrossing field of cryptology. Their contributions enriched it with dozens of new cipher systems.[34]

It is thanks to the simultaneous discovery of a radically important economic invention—the telegraph—and a realisation that the telegraph was unable to provide the sort of communications security needed to gain the full economic benefit of that invention that we have the modern field of cryptography. Subsequent communications revolutions have followed similar paths. When the telephone was first introduced in the first decades of the twentieth century, switchboard operators matched callers together and, while they were forbidden from listening in, could do so easily if they wished—another stable of popular culture. Automatic switchboards provided a leap in communications privacy. Likewise, the prevalence of "party lines"—telephone lines shared between groups of subscribers—allowed neighbours to listen in to conversations as they saw fit. One of the major drivers of the demand for private telephone lines was the possibilities of providing medical

34 Ibid.

advice over the phone, particularly in rural areas.

Digital messaging at first seemed to follow this trajectory. Early email communication tended to use shared terminals—which were accessible to colleagues—and open discussion lists. Over time the introduction of specific user-access controls and personal email accounts protected behind a password became dominant, building a degree of privacy into this increasingly important form of communication. End-to-end encrypted email systems began to develop throughout the 1970s and 1980s. Yet fully private email has never gained a significant market share. While an individual user might hide their emails behind passwords and pin numbers, it is typically the case that employers and email service providers have relatively free access (although sometimes limited by both law and practice) to the contents of those email inboxes.

It is only in recent years that lack of privacy over personal digital communication has begun to change. Email communication is deeply embedded in the business world yet there has been little innovation in email. The email experience of the 1990s is little different to that of the 2020s. Yet the ubiquity of smartphones in the 2010s has led to the rapid spread of instant messaging and chat services which, as this chapter has discussed, have responded to market demand for greater levels of privacy—both from the surveillance of the state and the surveillance of the workplace.

In this way, the history and evolution of privacy offers an important lesson for the history of freedom in general. For the most part, privacy successes have not been won through the development of regulation—they have not been won in court. While privacy advocates focus their attention on the domain of politics and legislation, it is in privacy-preserving innovation that we ought to focus our attention.

These innovations come partly from market demand—the demand by patients to communicate with their doctors via telephone, of firms to share information without the fear of being pried on—but also by independent innovators who are willing to offer consumers something that consumers may not know they want. The role of liberty-focused entrepreneurs and innovators isn't just to convince people that privacy is important, but to build them the tools to protect it themselves.

07.

SPECIAL JURISDICTIONS AND CRYPTODEMOCRACIES

Political reform is hard and slow.[1] Political reform in favour of liberty is particularly difficult because political elites often have strong incentives to maintain illiberal policies.[2] In this chapter we explore an alternative strategy to expand our freedom: build new jurisdictions. These "special jurisdictions"—free ports, private cities,

1 This chapter draws on a book and two papers with Aaron M Lane and Jason Potts: Allen, Darcy W.E., and Aaron M. Lane. 2020. "Cryptodemocratic Governance in Special Economic Zones." *Journal of Special Jurisdictions* 1 (1); Allen, Darcy W.E., Chris Berg, and Aaron M. Lane. 2019. *Cryptodemocracy: How Blockchain Can Radically Expand Democratic Choice*, edited by Lenore T. Ealy and Paul Aligica. Lanham, US: Lexington Books; Allen, Darcy W.E., Chris Berg, Aaron M. Lane, and Jason Potts. 2018. "Cryptodemocracy and Its Institutional Possibilities." *The Review of Austrian Economics*: pp. 1-12.

2 See, for example: Acemoglu, Daron, and James A. Robinson. 2000. "Political Losers as a Barrier to Economic Development." *American Economic Review* 90 (2): pp. 126-130.

special economic zones—are a longstanding mechanism used by states to obtain greater economic freedom for business. Special jurisdictions are now more viable and more adaptable under the wave of technological innovation we have explored in this book. As with the nation state, special jurisdictions need governance. Blockchain technology offers a new mechanism to govern these jurisdictions, and a new way to surmount some of the long-standing challenges with realising the opportunities of competitive innovative governance.

Special jurisdictions are small geographical areas where there are different regulatory rules (e.g. lower taxes, fewer regulations, different legal systems) compared to the jurisdiction that hosts them.[3] For example, a state might carve out a region and exclude businesses that establish themselves in that region from paying corporate tax, or being subject to minimum wage rules. Or a city might nominate a few blocks where firms do not have to pay property taxes.

There are now thousands of special jurisdictions globally. They differ in terms of their policies, size, funding and governance. But all these new jurisdictions represent a decentralisation of political power—from central political authorities to regional and local ones.[4] Special jurisdictions have historically generated political, social and economic freedom. This shift is a process of greater institutional diversity, choice

---

3 Akinci, Gokhan, and James Crittle. 2008. "Special Economic Zones: Performance, Lessons Learned, and Implications for Zone Development." Washington, DC: The World Bank Group. We also include sea steading as an example of institutional competition—an extreme form of special zones that don't rely on permission from government (at least in theory).

4 The World Bank notes some general principles of a special economic zone including a "geographically delimited area", "single management/administration," "eligibility for benefits based on physical location within the zone." See ibid., p. 9.

and competition. Just as we desire competition in products and services in markets, special jurisdictions give us greater competition over the rules that govern us.

The benefits of a special jurisdiction don't only accrue to the people living within that border, or even those people and businesses that migrate into new zones. Special jurisdictions can spark "liberalization avalanches" where more effective rules spread through the host jurisdiction and to neighbouring states. In China, for instance, the first special zone, Shenzhen, was implemented in 1980[5] In 2020 more Chinese people live within special zones than outside them.

We understand that special jurisdictions can be useful—they are tools of freedom—but why don't we have more of them? These innovative forms of governance pose complex and sometimes insurmountable governance problems. They don't just appear and evolve on their own—they must be instigated and implemented. These problems begin with the very process of founding a zone. Where should a zone be located, and how close is it to existing entrenched interests? On what policy margins should a zone differ from the host jurisdiction? How should the infrastructure be paid for—and what infrastructure should be built?

The laws and regulations governing special jurisdictions must also be able to evolve with economic and political conditions. What policies should change, and how should regulatory change occur? How can

---

5 Montinola, Gabriella, Yingyi Qian, and Barry R Weingast. 1995. "Federalism, Chinese Style: The Political Basis for Economic Success in China." *World Politics* 48 (1); Crane, George T. 1990. *The Political Economy of China's Special Economic Zones*. Routledge; Crane, George T. 1994. "'Special Things in Special Ways': National Economic Identity and China's Special Economic Zones." *The Australian Journal of Chinese Affairs* (32); Yeung, Yue-man, Joanna Lee, and Gordon Kee. 2009. "China's Special Economic Zones at 30." *Eurasian Geography and Economics* 50 (2): pp. 222– 240. p. 225.

special jurisdictions maintain governance autonomy from their hosts? How can administrative and government bodies within new jurisdictions commit to economic freedom, ensuring they won't extract rents later?

Solving governance challenges in special jurisdictions will not only make existing zones more effective, but also lower the costs of making more jurisdictions—enabling more experiments in freedom. In this chapter we examine how other new technologies can help to ameliorate some of these challenges. Blockchain provides a digital infrastructure for governing property rights in special jurisdictions and for making decisions about governance and institutions. What we call *cryptodemocracy*—the radical possibilities offered by blockchain for voting and collective choice—offers a way to help overcome the challenges of governing a special jurisdiction, generating more jurisdictional competition, and, ultimately, more freedom.[6]

The economist Albert O. Hirschman famously described two strategies that individuals can use to respond when an organisation's services get worse: voice or exit.[7] To use voice is to complain or convince others to complain. In the case of a private company, this might involve writing letters to management expressing frustration with lower quality goods. For citizens opposing their government, voice might involve protest or public agitation, with the goal of shifting public opinion in a way that the government must respond. Democracy is a

---

6 See: Allen et al., "Cryptodemocracy and Its Institutional Possibilities"; Allen, Berg, and Lane, *Cryptodemocracy: How Blockchain Can Radically Expand Democratic Choice*.

7 Economist Albert O. Hirschman distinguished between two different tools to achieve institutional or organizational change. See: Hirschman, Albert O. 1970. *Exit, Voice, and Loyalty: Responses to Decline in Firms, Organizations, and States*, vol. 25. Cambridge, MA: Harvard University Press.

way to embed the voice into the system of government. But regardless of whether a country is free or unfree, voice is a difficult strategy. Change takes time. Often political inertia is by design, and liberty-preserving. But much of what makes voice hard is the buildup of rent-seeking behavior by deals between the political and the business world.[8]

Where the voice strategy seeks to make a relationship better, the exit strategy involves leaving that relationship (usually with the intention of forging a new one). If product quality declines, we might move our business to another firm. Or if a government becomes repressive, we can try to physically migrate across borders to live under a different government. The ability for citizens to leave sparks a competitive process which creates incentives for governments to introduce better, less oppressive rules.

Exit also reveals information about what people want.[9] Indeed, as

---

8 We implement constitutional constraints or supermajority requirements to prevent tyranny of the majority, but this also has the effect of slowing institutional change. On the buildup of political rents, from the perspective of public choice, see: Buchanan, James M., and Gordon Tullock. 1962. *The Calculus of Consent*, vol. 3. Ann Arbor, MI: University of Michigan Press. On the economics of regulation (i.e. the demand and supply of regulation) see: Peltzman, Sam. 1976. "Toward a More General Theory of Regulation." In *National Bureau of Economic Research Working Paper 133*; Stigler, George J. 1971. "The Theory of Economic Regulation." *The Bell Journal of Economics and Management Science* 2 (1): pp. 3-21.

9 See: Buchanan, James M., and Viktor J. Vanberg. 1991. "The Market as a Creative Process." *Economics and Philosophy* 7 (2): pp. 167-186; Vanberg, Viktor, and Wolfgang Kerber. 1994. "Institutional Competition among Jurisdictions: An Evolutionary Approach." *Constitutional Political Economy* 5 (2): pp. 193-219. This process is important because we can never objectively know what an effective set of rules are. See, for instance: Allen, Darcy W.E., and Chris Berg. 2017. "Subjective Political Economy." *New Perspectives on Political Economy* 13 (1-2): pp. 19-40. See: Snow, Marcellus S. 2002. "Competition as a Discovery Procedure." *Quarterly Journal of Austrian Economics* 5 (3): pp. 9–23.

Viktor Vanberg and Wolfgang Kerber write, "institutional competition ought to be viewed as a knowledge-creating discovery process."[10] Political entrepreneurs and citizens coordinate plans and reveal preferences as an innovative process:

> the process of competition among jurisdictions can be seen, in analogy to market competition, as a process of experimenting, exploration, and discovery, in which alternative institutional arrangements or social technologies are tried out in an arena in which new arrangements and institutional inventions can constantly appear on stage, challenging established solutions.[11]

Jurisdictional competition exists today. Local governments, as Charles Tiebout famously pointed out, compete for the provision of public goods.[12] The states and territories compete with one another in federations. Nations compete on tax rates, regulatory environment and court systems.

Yet exit is costly. There are costs in physically moving and there are costs in identifying where to move. And of course, there are political restrictions on exit. The Berlin Wall was constructed to prevent East German citizens from exiting to West Germany in response to the low-quality institutions of the communist regime. Immigration restrictions prevent potential migrants from entering new jurisdictions. This

---

10 Vanberg and Kerber, "Institutional Competition among Jurisdictions: An Evolutionary Approach," At p. 216.

11 Ibid., p. 204.

12 Tiebout, Charles M. 1956. "A Pure Theory of Local Expenditures." *Journal of Political Economy* 64 (5): pp. 416-424.

means that in the real world institutional competition is limited. We can't easily decide to live under the rules of one jurisdiction one day, and some other jurisdiction the next, as we might shop at competing supermarkets when things are on sale.

So how can we lower exit costs? One obvious way to do this is through immigration reform: open the borders. Another possibility is to devolve political power to lower levels, including through the creation of new special jurisdictions. As Ilya Somin argues:

> The informational advantages of foot voting over ballot box voting strengthen the case for limiting and decentralizing government. The more decentralized government is, the more issues can be decided through foot voting. It is usually much easier to vote with your feet against a local government than a state government, and much easier to do it against a state than against the federal government.[13]

Special jurisdictions emerge for many reasons.[14] Governments might want to implement some liberty-enhancing reforms but believe they can't implement those reforms everywhere. In that sense new jurisdictions offer an "escape valve" for political deadlock. Governments might also be uncertain about the potential effectiveness of

---

13 Somin, Ilya. 2013. "Democracy and Political Ignorance." *Cato Unbound*, October 11.

14 For a more detailed analysis for some of the rationale behind special economic zones see: Madani, Dorsati. 1999. "A Review of the Role and Impact of Export Processing Zones." Washington, DC: The World Bank; Akinci and Crittle, "Special Economic Zones: Performance, Lessons Learned, and Implications for Zone Development."

reform. Special jurisdictions enable them to test reform—such as the implications of immigration restrictions or reducing licensing requirements—on a smaller scale. Then, if the reforms work, changes can be implemented more in the rest of the country.

But too often advocates of special jurisdictions assume that the political actors that introduce them are benevolent. Special jurisdictions have often been used to avoid liberalization, rather than to spur it.[15] Some special jurisdictions have been established in order to deliver preferential treatment to specific politically-connected firms. Special zones can also be used as a geopolitical weapon to establish dominance over other jurisdictions. Rome used special zones to punish nearby Rhodes. As trading activity shifted due to lower taxes, revenues in Rhodes fell.[16] More recently, Singapore was used as a kind of arbitrage between regulatory environments because private ships could circumvent the monopoly that the East India Company had over trade between India and China.[17]

What do special jurisdictions look like? A typical special jurisdiction is the Dubai International Financial Centre (DIFC), established in 2004.[18] The DIFC, which spans just 1.1 square kilometers, established

---

15 Moberg, Lotta. 2015. "The Political Economy of Special Economic Zones." *Journal of Institutional Economics* 11 (1): pp. 167-190.

16 Reger, Gary. 1994. *Regionalism and Change in the Economy of Independent Delos*. Berkeley, CA: University of California Press Berkeley; Gruen, Erich S. 1986. *The Hellenistic World and the Coming of Rome*, vol. 1. Berkeley, CA: University of California Press. p. 304.

17 Bernstein, William J. 2009. *A Splendid Exchange: How Trade Shaped the World.* Grove Press. p. 292.

18 Interestingly the creation of this zone itself was the product of a clear devolution of power between the emirates within the United Arab Emirates, freeing Dubai to make such radical changes.

its own regulatory environment and court system through common law institutions, including hiring a British Judge. The zone also relaxed ownership laws (such as having domestic firm ownership), implemented a 40-year tax-free guarantee, and provided arbitration services.[19] One of the problems with small zones such as a DIFC is that they might limit the emergence of entire cities, and therefore inhibit the benefits of city life and exchange. While small zones can be effective, there is evidence that larger zones might be more effective.[20]

Of course, the success of a zone is highly dependent on the policies implemented in that zone. So-called "single factory zones" provide preferential rules to a single company and are arguably not special zones at all.[21] Highly discriminatory zones might even do more harm than good (particularly if the host economy is already liberalized) because they create more political opportunities for rent-seeking.[22] More broadly, some jurisdictions enable the warehousing, storage and distribution of goods before they are re-exported (with lower trade barriers such as tariffs), but maintain other restrictions (e.g. portions of production within the zone must be exported).

---

19 For a review see: Strong, Michael, and Robert Himber. 2009. "The Legal Autonomy of the Dubai International Financial Centre: A Scalable Strategy for Global Free-Market Reforms." *Economic Affairs* 29 (2): pp. 36-41.

20 "The size of the SEZ is positively and significantly correlated with zone performance." See: Frick, Susanne A., Andrés Rodríguez-Pose, and Michael D. Wong. 2019. "Toward Economically Dynamic Special Economic Zones in Emerging Countries." *Economic Geography* 95 (1): pp. 30-64. p. 49.

21 This could include, for instance, one category of production process.

22 "As a policy of differential taxation and trade restrictions, SEZs will then likely create rents, rather than to lower them." Moberg, Lotta, and Vlad Tarko. 2014. "Why No Chinese Miracle in Africa? Special Economic Zones and Liberalization Avalanches." In *Special Economic Zones and Liberalization Avalanches*. At p. 5.

Special jurisdictions evolve over time. In the 1934 Foreign Trade Zones Act in the United States created highly limited zones that maintained the same labor laws and regulations as the areas that they were carved from. In 1950, the Boggs Amendment loosened the restrictions on these zones, broadening their focus beyond simply manufacturing, spurring more free zones across the country. China's Shenzhen was initially focused on high-technology firms, but this direction was later relaxed, attracting a significant amount of investment from different types of manufacturing, including textiles.

Another evolution has been the shift of how special jurisdictions are funded and governed. Over the twentieth century governments handed much responsibility of special jurisdictions to non-government authorities.[23] A recent stock take of 2,301 zones in the developing world found that 62 per cent were developed and operated by the private sector.[24] Privately operated zones account for around $200 billion in gross exports each year, and are responsible for the direct employment of over 40 million people.[25] These private zones are "less expensive to develop and operate… and yield better economic results," according to one assessment.[26]

SEASTEADING AS SPECIAL JURISDICTIONS

While the private sector might run a special jurisdiction, these areas are typically carved out of an existing political regime as a matter of

---

23 Akinci and Crittle, "Special Economic Zones: Performance, Lessons Learned, and Implications for Zone Development."

24 Ibid.

25 Ibid., p. 23.

26 Ibid., p. 4.

deliberate government policy. A special economic zone, or free trade area, is a policy decision—it is initiated and approved by the state. In that case, voice must precede exit—seeking permission from the very political systems that we're trying to fix. This permissioning process constrains the extent of jurisdictional competition. An even more ambitious attempt by the private sector to create new special jurisdictions is through seasteading. Like the homesteaders from which they derive their name, seasteaders seek to occupy uninhabited (and ungoverned) international waters and create independent communities separate from the state. Seasteaders don't merely seek to exit a political system, but to create entirely new political systems on the 71 percent of the planet that is ocean: to create "permanent, autonomous settlements on the ocean."[27]

The idea of seasteading was developed in the 1980s and 1990s. Several writers and activists have sought to secede from the state by living permanent or semi-permanent existences on the ocean. Having spent years on the ocean, the sailor Ken Neumeyer wrote the 1981 book *Sailing the Farm*, which offered a guide on how to sustain oneself on the ocean: how to distill water for drinking, how to prepare and eat seaweed, and how to preserve and secure food purchased at ports. All this was to secure an independent liberty outside the reaches of government:

> The freedom to roam where intuition guides, to go where food is cheap or free and to live without that "big brother is watching you" feeling does not seem to be possible on land. Real estate tax, sales tax, income tax, Social Security tax, etc., etc. Cut it

---

27 Friedman, Patri, and Brad Taylor. 2012. "Seasteading: Competitive Governments on the Ocean." *Kyklos* 65 (2): pp. 217-235. At pp. 217-218.

> loose! Get a boat and head for the open sea.[28]

Likewise, Wayne Gramlich, author of an influential 1998 essay which proposed building a seasteading community using thousands of empty two litre plastic beverage bottles, wrote:

> Why would anyone want to colonize the ocean surface? There are a number of reasons — adventure, religious freedom, tax avoidance, trying out new forms of government, etc. Of the ones listed, tax avoidance is my pick as the most powerful motivator for the development of sea surface colonization technology.[29]

The seasteading community is a curious combination of libertarian, ecological and utopian values. In seasteading, environmentalism, recycling and ideas of self-sufficiency coexist with anti-state sentiments, enhanced with a strong vision of a radically different world. At its most ambitious, seasteading thought can cross over into science fiction. In his book *The Millennial Project*, the writer Marshall T. Savage spelled out a plan for independent colonization of the galaxy that begins with the experimental creation of colonies on the ocean. Spacefarers would be required to minimise the use of resources, grow their own food and limit their environmental impact. These were skills that Savage believed could first be mastered in seasteading: "Marine colonies, will, like space colonies, make use of space which is now ecologically barren. The open oceans are largely lifeless due to lack of nutrients.

---

28 Neumeyer, Ken. 1982. *Sailing the Farm.* Ten Speed Press.

29 Gramlich, Wayne. 1998. "Seasteading – Homesteading the High Seas."

The marine colonies will therefore displace no existing ecosystems."[30]

Actual seasteading projects have, however, been somewhat limited. We could describe the "pirate republics" documented by the economist Peter Leeson as a form of mobile seasteading, as communities of pirates denied the authority of sovereign states and established their own constitutional orders.[31] The 1960s and 1970s saw a wave of attempts to establish micronations.[32] One of the most famous progenitor projects, the Principality of Sealand, involved the occupation of an offshore platform in the North Sea that was originally built as an anti-aircraft fort called Roughs Tower. It is now claimed to be an independent nation, although is unrecognised by any other sovereign state—including the United Kingdom whose territorial waters it exists within.

Sealand, however, has been relatively lucky in that it has not been the subject of direct action by nearby governments. The Republic of Minerva, established on reefs in the South Pacific in 1972, collapsed when the nearby state of Tonga made it clear that it would not tolerate "people setting up empires on our doorstep."[33] The Republic of Rose Island was a 1968 project on a specially constructed platform in the

---

30 Cited in: Quirk, Joe, and Patri Friedman. 2017. *Seasteading: How Floating Nations Will Restore the Environment, Enrich the Poor, Cure the Sick, and Liberate Humanity from Politicians.* Simon and Schuster. At p. 30-31.

31 Leeson, Peter T. 2007. "An-Arrgh-Chy: The Law and Economics of Pirate Organization." *Journal of Political Economy* 115 (6): pp. 1049-1094; Leeson, Peter T. 2009. *The Invisible Hook: The Hidden Economics of Pirates.* USA: Princeton University Press.

32 Menefee, Samuel Pyeatt. 1994. "Republics of the Reefs: Nation-Building on the Continental Shelf and in the World's Oceans." *California Western International Law Journal* 25 (1).

33 Trumbull, Robert. 1972. "Pacific Islanders Fight Reef Plan." *The New York Times*, February 27.

Adriatic Sea, and was demolished by Italian police shortly after it declared its sovereignty.[34] A seastead off the coast of Phuket established in 2019 only lasted a couple of months before the Thai navy boarded it and charged its occupants with violating Thailand's sovereignty.[35]

As this history vividly shows, sovereign nations do not like the idea of institutional competition. Seasteads also awkwardly fit into existing international naval law and custom. If ships in international waters don't fly the flag of a terrestrial nation they are treated as "inherently suspect as presumptive pirate, smuggler, illegal fisher, or other enemy of civilization".[36] Many seasteads are therefore reliant on flying some flag of an existing jurisdiction. In their book *Seasteading: How Floating Nations Will Restore the Environment, Enrich the Poor, Cure the Sick, and Liberate Humanity from Politicians*, Joe Quirk and Patri Friedman argue that seasteads should be able to rely on the same legal rights and protections under maritime law as other floating homes, such as cruise ships.[37] Several major cruise liners fly 'flags of convenience'—such as those provided by Panama, the Bahamas and Libera. While the flags of convenience approach helps deal with some of the legal challenges,

---

34 Cerviere, Giacinto. 2009. "The Rose Island." *Abitare*; Hayward, Philip. 2014. "Islands and Micronationality." *Shima* 8 (1).

35 ABC News Australia. 2019. "Seasteading Bitcoin Couple Charged with Violating Thai Sovereignty as Navy Boards Floating Home." *ABC News Australia*, April 21.

36 Bell, Tom W. 2017. *Your Next Government? From the Nation State to Stateless Nations*. Cambridge: Cambridge University Press. p. 59.

37 Quirk and Friedman, *Seasteading: How Floating Nations Will Restore the Environment, Enrich the Poor, Cure the Sick, and Liberate Humanity from Politicians*.

the longer-term prospects of seasteading under this model are murky.[38]

The Seasteading Institute, established in 2008, has been working towards fixing the institutional framework governing seasteads by cooperating with existing governments. In 2017, it signed an informal agreement with the government of French Polynesia to set up seasteads in their waters. But this approach makes the seastead vulnerable to policy change in the host nation. Public protests in Tahiti in 2018 from fishers who objected to the location of the proposed seastead led to the government's withdrawal of support, declaring that the 2017 agreement was void.[39]

Seasteading brings into sharp relief the fundamental challenge with all special jurisdictions, whether they are created by sovereign governments or not: governance. Most arguments for special jurisdictions focus on how they influence the allocation of resources. Do special jurisdictions increase trade or attract foreign investment? But more elemental for their success is their governance.[40] How do they relate to other jurisdictions—other states? How can they manage themselves?

THE SPECIAL JURISDICTION GOVERNANCE PROBLEM

The economist Lotta Moberg has argued that for a special jurisdiction to be successful, "decision makers need both be able to find the

---

38 For more on the problems with "flags of convenience" see: Neff, Robert. 2007. "Flags That Hide the Dirty Truth." *Asia Times*, April 19.

39 Radio New Zealand. 2018. "Hundreds March in Tahiti Against Building of Floating Islands." *Radio New Zealand*, April 9.

40 For example, see: Aggarwal, Aradhna. 2007. "Impact of Special Economic Zones on Employment, Poverty and Human Development." *Indian Council for Research on International Economic Relations (ICRIER) Working Paper No. 194*; Wang, Jin. 2013. "The Economic Impact of Special Economic Zones: Evidence from Chinese Municipalities." *Journal of Development Economics* 101 (c).

proper policies for the zones and have the incentive to implement them."[41] Special jurisdictions come into the world as either the result of government policy—top-down directives from a large state—or as a rebellious colonisation of unoccupied territory at sea. At a first glance, neither situations are conducive to good governance. Special economic zones are vulnerable to being captured by rent-seekers. The precarious existence of seasteads make them vulnerable to predatory states. But as Moberg points out, to be prosperous, special jurisdictions need the same characteristics as successful states: democracy, political decentralisation, and private market economies.[42]

Friedrich Hayek famously argued that information is distributed across the minds of individuals—not centrally organised—and is often competing and contradictory.[43] This "knowledge problem" is critical in an economy, and one of the reasons that market systems outperform socialist systems, but it is particularly severe in special jurisdictions, where not only the distribution of resources is unknown but also what sort of institutions should be designed.[44] For special jurisdictions the rules of the game themselves need to be discovered.

Even seemingly simple decisions such as the geographical location

---

41 Moberg, "The Political Economy of Special Economic Zones," p. 4.

42 Ibid.

43 Hayek, Friedrich A. 1945. "The Use of Knowledge in Society." *The American Economic Review* 35 (4); Hayek, Friedrich A. 1937. "Economics and Knowledge." *Economica* 4 (13).

44 For example, see: Bylund, Per L., and Matthew McCaffrey. 2017. "A Theory of Entrepreneurship and Institutional Uncertainty." *Journal of Business Venturing* 32 (5). On institutional entrepreneurship in a development context see: Leeson, Peter T., and Peter J. Boettke. 2009. "Two-Tiered Entrepreneurship and Economic Development." *International Review of Law and Economics* 29 (3).

of a zone can make the difference between success and failure.[45] What tax or regulator concessions should the jurisdiction offer? Who should decide? What decisions should be made by the private sector and what are made by the government?[46] What happens when the interests of the host jurisdiction and the special jurisdiction diverge? Special jurisdictions also need mechanisms to adapt—to learn and evolve as new information comes to light. A jurisdiction that might start by reducing tariffs might also need to make concessions on immigration policy to attract workers.

For example, at the tail end of the Cold War the Soviet Union established special economic zones in an unsuccessful attempt to replicate the success of China's Shenzhen. One of the most significant was Nakhodka, a port in the far eastern region of Vladivostok, designated as a special economic zone in 1990. The site was ideal: Nakhodka is a strategically located trading port between Russia, Japan and Korea that also serves as a Russian gateway to the United States and the rest of Asia. But the zone was hampered from the start by institutional rigidities—bureaucratic inertia, and a reluctance to devolve power to the zone meant that there was little Nakhodka could do to thrive

---

45 Choosing a zone location is complex in part because it must consider a complex mix of existing entrenched interests within the host jurisdiction.

46 In this sense SEZs require an analysis through the lens of entangled political economy, where there is a co-evolution of the private and public spheres of activity. See: Wagner, Richard E. 2016. *Politics as a Peculiar Business: Insights from a Theory of Entangled Political Economy.* Cheltenham, UK: Edward Elgar Publishing.

among the Soviet Union's general collapse.[47]

Obviously, privatisation can help to align the incentives of investors, the governance structures and the citizens by giving the governing body a stake in the outcome.[48] But the idea of privately governed cities or jurisdictions is at first confusing. Who would provide the roads, sewerage and electricity connection? How would disputes be resolved without courts? Where do basic services such as police and a fire service come from?

But think of your local shopping centre: the owner of the centre collects rents and provides infrastructure and services such as cleaning, security and other amenities. The owner of the property rights (of the city, zone, shopping center) has the incentive to maximise the value of those property rights. Because they are the residual claimants on the profits of that land, they have the incentive to keep residents happy by providing physical infrastructure, such as roads, and institutional infrastructure, such as courts and governance. Because a (potentially large) driver of the value of land is the quality of governance coupled with a private residual claimant means that the proprietor might be incentivised to provide high quality institutional infrastructure. In the same way, private cities potentially minimise rent seeking by making those making decisions the residual claimants of those decisions. This encourages decision makers to provide infrastructure and services.

---

47 Miller, Chris. 2016. *The Struggle to Save the Soviet Economy: Mikhail Gorbachev and the Collapse of the USSR.* US: University of North Carolina Press Books. Vikhoreva, Svetlana J. 2001. "The Development of Free Economic Zones in Russia." *The Economic Research Institute for Northeast Asia (ERINA) Report* 38; Manezhev, Sergei. 1993. "Free Economic Zones in the Context of Economic Changes in Russia." *Europe-Asia Studies* 45 (4).

48 Lutter, Mark. 2017. "The Case for Innovative Governance." *Centre for Innovative Governance Research*; Lutter, Mark. 2016. "Three Essays on Proprietary Cities." *George Mason University.*

One further benefit of privatisation of governance is the potential to enact wholesale reform.

Defining private cities is difficult because the lines of private and public are blurry (particularly with the rise in public-private zones where some infrastructure is paid for by the private sector and other parts by the public sector). According to Mark Lutter, the head of the Center for Innovative Governance, proprietary cities have three main characteristics: the land owner as a private, for-profit entity; they have a high degree of legal and regulatory autonomy; and they have a "meaningful role in creating and enforcing the legal system."[49] The role of the host government in a proprietary city is to provide autonomy to a separate company or group of companies to govern over a territory.[50]

But public private partnerships (PPP) and private cities means an increase in the number of stakeholders in relation to the zone. Governance must achieve coordination of a wide range of stakeholders: investors being residual claimants to gain rents and align incentives; citizens to make the rules legitimate; and host governments to ensure that their commitments to autonomy are credible. Given the perhaps radical institutional and policy changes that happen in an SEZ—that is necessary and desired from the perspective of solving knowledge problems—they need a collective choice infrastructure to coordinate preferences.

Democratic governance and accountability is another important check on the challenge of opportunism. Those who govern should be accountable for their decisions, and therefore gain from making good choices through retaining governance power. Coupled with this is the

49 See "Three Essays on Proprietary Cities," 2.

50 Ibid.

notion of decentralization of governance that gives local officials the incentive to (try to) make good choices.[51]

Jurisdictional competition requires governance autonomy. Autonomy facilitates experimentation and can help to prevent encroachment of entrenched interests. But devolving power to smaller jurisdictions also creates internal problems (such as accountability and responsiveness to the people and companies within the zone) and external governance problems relating to other jurisdictions (such as maintaining autonomy and diplomatic ties). Special jurisdictions can have tumultuous relationships with host governments. For instance, in 2011 the Honduran Congress (almost unanimously) voted in favour of constitutional amendment to enable a new form of special zones.[52] These zones were to be extensive including more autonomous commercial laws, public administration, courts and policy—including the power to tax. In 2012, the Supreme Court struck down the proposal on the basis that it was unconstitutional. Later, in 2013, Honduras backflipped and passed legislation that enabled another type of special zone called a ZEDE. This regime uncertainty discourages investment.

Autonomy can be enhanced by other mechanisms that align incentives between special jurisdictions and their hosts. For instance, special jurisdictions can remit money back to the central government but remain the residual claimant on profits. In Honduras, for instance, each ZEDE is required to remit 12 percent of tax revenues back to the central government—this gives host governments a stake in the success of the lower-level jurisdiction.

---

51 Moberg, "The Political Economy of Special Economic Zones," p. 20.

52 For discussion see: Bell, *Your Next Government? From the Nation State to Stateless Nations*.

Even if greater governance autonomy is achieved, special jurisdictions must also have collective choice infrastructure for accountability to people within the zone (that is, they need their own collective action governance mechanisms to ensure responsiveness and representation from the people). We can now turn to how new technologies might help to solve some of these governance challenges.

Coordinating decisions in small homogenous groups is easy. With more participants and with diverse preferences, group decisions become hard. There are substantial transaction costs in integrating preferences into decisions about what governments should, shouldn't, can, and cannot do. Typically, liberal modern states use democracy to make these decisions. As Hayek wrote, the benefits of democracy come through its "dynamic process" and the "formation of opinion":

> Democracy is, above all, a process of forming opinion. Its chief advantage lies not in its method of selecting those who govern but in the fact that, because a great part of the population takes an active part in the formation of opinion, a correspondingly wide range of persons is available from which to select... It is in its dynamic, rather than in its static, aspects that the value of democracy proves itself. As is true of liberty, the benefits of democracy will show themselves only in the long run, while its more immediate achievements may well be inferior to those of other forms of government.[53]

Yet democracy has deficiencies. The economist Bryan Caplan has influentially pointed to the problems of voter ignorance and irrationality, noting that democracy offers little incentive for voters to acquire

---

53 Hayek, *The Constitution of Liberty: The Definitive Edition*, 174.

additional knowledge about political matters. Because there is only a tiny chance that any one vote will have any decisive impact on the outcome of an election, political ignorance is rational.[54] Various solutions have been proposed to the problem of voter ignorance, but most drawn from the liberty movement have focused either on shrinking the domain of policy questions that are subject to democratic decision-making, or somehow weighing the votes of more informed voters higher than ignorant ones.[55]

Caplan points out that "in democracies the main alternative to majority rule is not dictatorship, but markets. A better understanding of voter irrationality advises us to rely less on democracy and more on the market."[56] But political decision-making is inevitable in any community that needs to make policy decisions. At the first instance, the community has to decide what questions to leave to the market, or just decide who can make that choice. For special jurisdictions, these questions are troublesome. Typically special jurisdictions are established with a broad policy framework in mind—a special economic zone might be designed to have lower taxes, or a seastead might be designed to escape an oppressive national sovereign—but the need for policy change in response to changing circumstances creates significant

---

54 Downs, Anthony. 1957. "An Economic Theory of Political Action in a Democracy." *The Journal of Political Economy* 65 (2); Caplan, Bryan. 2011. *The Myth of the Rational Voter: Why Democracies Choose Bad Policies*. New Jersey, US: Princeton University Press.

55 For the former see: Somin, Ilya. 2016. *Democracy and Political Ignorance: Why Smaller Government Is Smarter*. Stanford University Press. For the latter see: Caplan, *The Myth of the Rational Voter: Why Democracies Choose Bad Policies*; Brennan, Jason. 2016. *Against Democracy*. New Jersey, US: Princeton University Press.

56 Caplan, *The Myth of the Rational Voter: Why Democracies Choose Bad Policies*.

frictions.

It is in coordinating large groups with diverse preferences that blockchain offers new possibilities in the organisation of democratic choice. In work with Aaron Lane, we have described a democratic system called "cryptodemocracy."[57] The fundamental idea behind cryptodemocracy is the use of blockchain to provide citizens with stronger property rights over their own vote—and in doing so create more effective governance infrastructure.

The idea that an individual owns their vote is intuitive—in that we as voters have the right to use our vote as we see fit. But our "right" to vote is highly regulated. While each democratic system is different, they tend to be similarly constrained. We can only cast our vote for our representatives at set times (perhaps every three to five years). We can only cast our vote for candidates within the geographical area we are registered. We have to make our vote in secret at a specified polling place (in the sense your vote cannot be externally verified). We cannot withdraw our vote once it is cast (wait three more years). We cannot sell or delegate our vote (that is, our vote is not alienable). We vote almost always for representatives to make decisions on our behalf, not on specific policy issues themselves. Some nations with compulsory voting, such as Australia, even force us to vote (or risk being fined).

Under a cryptodemocratic system, voters have full property rights over their vote to use as they see fit. Each voter is given a token, recorded on a blockchain, that represents their vote. They can then contract those votes to others using smart contracts—including

57 Allen, Berg, and Lane, *Cryptodemocracy: How Blockchain Can Radically Expand Democratic Choice*; Allen et al., "Cryptodemocracy and Its Institutional Possibilities."

imposing conditions on how those votes might be used—with the blockchain ledger providing a secure record who owns what. Those votes might be further delegated, brought or sold, perhaps in return for money or services. The result is the emergence of new polycentric forms of political power, driven by individual freedom in voting property rights. Together, blockchains and smart contracts enable votes to become programmable and contractable.

While the exact scope and application of these cryptodemocracies will be discovered over time, we can explore some of the features of cryptodemocratic governance rather than more conventional governance structures. Through the delegation (and buying and selling) of votes, we would anticipate a cryptodemocracy to have interesting properties.

First, cryptodemocracies will have emergent structures and centres of decision making. That is, cryptodemocracies will be polycentric. Cryptodemocracies are hard to define precisely because their structures are a "cosmos" (a spontaneously emergent order) rather than a "taxis" (a consciously planned order). The constellation of voting property rights at any given time is a function of voter preferences. For instance, politically active and aware voters could maintain their voting property rights and exercise the votes themselves. Others will delegate their entire voting rights for long periods of time to others. This also suggests that the stability of a cryptodemocracy is a result of individual preferences for that stability (e.g. delegating votes for longer time periods).

Second, cryptodemocracies will be more knowledge-rich than conventional democratic structures. The process of delegation and coordination between voters and delegates will integrate more local and contextual knowledge into collective choices. Some of that knowledge might come through the buying and selling of vodting rights—which enables people to demonstrate their intensity of preference.

Other knowledge will be integrated into the system through the process of selective delegation and the conditional breakdown of votes. Individuals have more power to self-identify how and in what way they wish to take part in the political process. It's worth examining again here some of the claims of voter ignorance and irrationality as described previously. In a cryptodemocracy voters—by having property rights within their own vote—will not homogeneously have the right to exercise their vote within predetermined constraints. Indeed, rather than some homogenous "bundle" of voter property rights being defined, individuals can more fully determine how they wish to engage.

Now that we have suggested some of the characteristics of a cryptodemocracy—that it is emergent, polycentric, knowledge-rich and bespoke—we can explore how this new type of collective choice infrastructure might ameliorate some of problems of special jurisdictions.

Special zones are a complex and entangled mix of public and private spheres of governance.[58] Cryptodemocratic governance can be applied to solve some of the governance problems underpinning special jurisdictions. There is potential here both for public elections (e.g. electing governments or administrative bodies) as well as within and between private and not-for-profit organisations (e.g. corporate shareholder voting or union governance). These opportunities could ameliorate some of the knowledge problems of special zones by incentivising the revelation of preferences and by making decision making more dynamic.

Cryptodemocratic governance may also create new incentive structures, bringing about more responsive and accountable governance structures. Together with privacy technologies and smart contracts

---

58 On entangled political economy see: Wagner, Richard E. "The Peculiar Business of Politics." *GMU Working Paper in Economics No. 16-27.*

outlined in previous chapters we may also be able to make zones more censorship-resistant to host nation demands. Smart contracts could be used to overcome the hold-up problem, disentangling different political systems and ensuring ongoing credible commitments around the terms of jurisdictional autonomy.

As we saw previously, a fundamental problem for special jurisdictions is that they must discover new policies. The epistemic properties of cryptodemocratic governance—through the delegation and unbundling of rights—might enable more knowledge to be integrated into those political decisions, propelling a further process of institutional jurisdictional competition. In this way cryptodemocracies might provide the collective choice infrastructure for better public governance within special jurisdictions, lowering the costs and increasing the capacity of zone governance.

The applications of cryptodemocracies extend into the private sector too. As we have seen, there are an increasing number of zones that are privately funded and operated, and some data suggests that private zones are more effective. One clear application of cryptodemocratic governance is for corporate shareholder voting. Cryptodemocracies can be used to facilitate democratic processes within joint stock companies. Shareholders today are already contracting and delegating voting rights. Cryptodemocratic governance might make those processes much more efficient through blockchain-enabled share registries. This might further facilitate the development of privately funded and operated special economic zones. More extreme, SEZs could use Distributed Autonomous Organisations (DAOs) that draw on the local knowledge of citizens (or some other franchise or investor group) to make decisions around funding of future infrastructure projects.

We can't just advocate for more special jurisdictions and expect our freedoms to expand. The economic and social infrastructure of those new jurisdictions also needs to be entrepreneurially built—and it needs to be built with freedom in mind. We argue that special jurisdictions, as testbeds for new policies, are an opportunity for experiments in freedom with frontier technologies. Particularly in greenfields sites, special jurisdictions are often comparatively free from the entrenched interests and restrictive regulatory environments of established states. The potential for cryptodemocratic governance we have described here is just one example for liberty-enhancing entrepreneurship within special jurisdictions. Many of the new modes of economic and social organisation we have explored in this book—from private social insurance mechanisms to the layered legal systems we turn to in the following chapter—could first be trialled within these radical forms of innovative governance.

08.

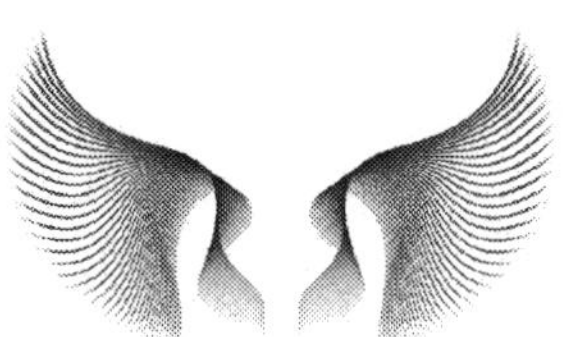

TECHNOLOGY FOR BETTER INSTITUTIONS IN POOR COUNTRIES

As the economy becomes more digitised, a greater share of the world's economic activity becomes more mobile and therefore easier to relocate in jurisdictions with better institutions.[1] Inhabitants of a twenty-first century seastead or special jurisdiction can be connected seamlessly to the global digital economy at relatively low cost. At the same time, the history of special jurisdictions shows that they can exert a surprising influence over the global economy themselves, providing opportunities and exporting services

1 This chapter draws on a working paper (with Jason Potts): Allen, Darcy W.E., Chris Berg, and Jason Potts. 2019. "Blockchain Technology and the Theory of Economic Development." *SSRN Paper No. 3333568.*

to people who remain within traditional state organisations.

Shenzhen was a magnet for poorly paid Chinese citizens who rushed into the growing city attracted to the market-based employment system. Shenzhen's economic success spilled over into the rest of China—not simply providing a model that could be replicated, but attracting investment that benefited other areas of the country as well.[2]

Likewise, many of the early seasteads hosted pirate radio stations—clandestine radio stations that broadcast without obtaining radio spectrum licenses. [3] Before it was established as Sealand, Roughs Tower was part of a network of sea-born ships and towers that supported the vibrant pirate radio community of the 1960s, as entrepreneurs challenged the staid monopoly of the British Broadcasting Corporation.[4] Pirate radio stations took advantage of the borderless nature of radiofrequency spectrum to subvert regulatory controls and national borders. More consequential were the radio stations that broadcast into the Soviet bloc during the Cold War; the cat-and-mouse game of radio jamming and counter-jamming by state and independent broadcasters distributed uncensored information to citizens whose governments maintained strict limits on free speech.[5]

---

2 Du, Juan. 2020. *The Shenzhen Experiment: The Story of China's Instant City.* US: Harvard University Press.

3 Walker, Jesse. 2004. *Rebels on the Air: An Alternative History of Radio in America.* New York, US: New York University Press.

4 Urbina, Ian. 2019. *The Outlaw Ocean: Journeys across the Last Untamed Frontier.* Knopf Doubleday Publishing Group; Johns, Adrian. 2010. *Death of a Pirate: British Radio and the Making of the Information Age.* WW Norton & Company.

5 Woodard, George W. 2010. "Cold War Radio Jamming." In *Cold War Broadcasting: Impact on the Soviet Union and Eastern Europe, a Collection of Studies and Documents*, edited by Ross Johnson and Gene Parta. Budapest and New York: CEU Press.

In this way, the citizens of one jurisdiction can, through their own efforts, materially make the citizens of another jurisdiction freer without the former having to cross physical geographic borders. We have already seen how entrepreneurs in the West have developed technologies and mechanisms for Chinese citizens to subvert the Great Firewall of China. But technologies that cross borders, providing a mechanism for individual freedom, turns out to have even more radical implications.

In his book *The Political Economy of Non-Territorial Exit*, Trent MacDonald describes how blockchain technology can be used as a mechanism for citizens to "exit" undesirable or poorly functioning state institutions without having to cross political borders. What MacDonald calls "cryptosecession," describes:

> the phenomenon of individuals seceding from state-run institutions (and thus jurisdictions) not by physically leaving, but by using cryptographic technologies such as Bitcoin and other blockchain applications to exit to virtual "states."[6]

Rather than having to physically exit the jurisdiction of the nation state, people can live within the same geographical borders but non-territorially exit to blockchain-based institutions. The benefit to citizens here is clear: they can now opt-in to a wide range of different institutions of money, property rights, contract enforcement, and so on,

---

6 MacDonald, Trent J. 2019. *The Political Economy of Non-Territorial Exit*. Edward Elgar Publishing; MacDonald, Trent J., Darcy W.E. Allen, and Jason Potts. 2016. "Blockchains and the Boundaries of Self-Organized Economies: Predictions for the Future of Banking." In *Banking Beyond Banks and Money: A Guide to Banking Services in the Twenty-First Century*, edited by Paolo Tasca, et al. Springer International Publishing.

or remain within existing institutional systems. Mark S. Miller and Marc Stiegler describe this as the "digital path … made possible by new technologies that could leverage the existing wide recognition of first world institutions to short circuit the slow growth process of the other paths."[7]

But for citizens to cryptosecede they need alternative institutions to cryptosecede to. Behind this cryptosecession process is a background of institutional entrepreneurship (or "cryptostatecraft") where entrepreneurs build blockchain-based networked institutions.[8] There are many ways that we can conceive of this entrepreneurial process. It is distinct from more common forms of entrepreneurial discovery because it involves "protective-tier" entrepreneurship—searching for institutions that protect property rights. We might alternatively view this process as a type of "evasive entrepreneurship," or "the expenditure of resources and efforts in evading the legal system or in avoiding the unproductive activities of other agents."[9] Entrepreneurs can now invest resources in developing digital infrastructure that facilitates exchange by others.

Importantly this development process can happen without displacing existing state-based geographical institutions. The process of creating new institutions is messy. It involves many competing institutional

---

7 Miller, Mark S., and Marc Stiegler. 2003. "The Digital Path: Smart Contracts and the Third World." In *Markets, Information and Communication*, edited by Jack Birner and Pierre Garrouste. Routledge.

8 The term "cryptostatecraft" was coined by MacDonald.

9 Coyne, Christopher J. and Peter T. Leeson. 2004. "The Plight of Underdeveloped Countries." *Cato Journal* 24 (3): 235-249. See also Ekert, Niklas, and Magnus Henrekson. 2016. "Evasive entrepreneurship." *Small Business Economics* 47 (1): 95-113.

systems engendering an evolutionary approach to change. Blockchains facilitate that process—lowering the costs of developing secure, decentralised digital infrastructure. Institutional change is no longer restricted to change across entire political jurisdictions. Institutional change is expanded to a complex competing mix of private institutions built on top of existing ones, allowing individuals to innovate and adapt from within and without their own territory. And it is here that cryptosecession provides an enormous opportunity for the developing world.

THE PROBLEM OF ECONOMIC DEVELOPMENT

There are many economic theories that purport to explain why some countries are rich and others are poor. One theory suggests that developing economies are stuck in a poverty trap through lack of savings reducing investment and capital accumulation.[10] Other explanations suggest problems of structural adjustment between sectors, such as agriculture and manufacturing.[11] Or poor countries might have misallocated their resources because of few incentives to invest in

10 Early development economists proposed a *savings-investment gap* where developing economies had insufficient domestic savings necessary to satisfy levels of investment to build the capital stock. This gap left developing economies in a persistent stage of poverty. Many causes for such a savings shortfall and trap were suggested, including subsistence lifestyles, population growth, and some capital threshold necessary for additional productivity gains. See: Domar, Evsey D. 1946. "Capital Expansion, Rate of Growth, and Employment." *Econometrica, Journal of the Econometric Society* 14 (2); Harrod, Roy F. 1939. "An Essay in Dynamic Theory." *The Economic Journal* 49 (193); Rostow, Walt Whitman. 1990. *The Stages of Economic Growth: A Non-Communist Manifesto.* Cambridge University Press; Swan, Trevor W. 1956. "Economic Growth and Capital Accumulation." *Economic Record* 32 (2).

11 For instance: Lewis, W. Arthur. 1954. "Economic Development with Unlimited Supplies of Labour." *The Manchester School* 22 (2); Rostow, *The Stages of Economic Growth: A Non-Communist Manifesto.*

complementary capital stock.[12]

Each of these explanations for poverty suggest ways that public policy could reduce poverty. If there is a gap between savings and investment, targeted foreign aid could bridge the gap. That aid could come in the form of money, or donors could directly invest in infrastructure in poor countries.[13] Advocates of the "big push" approach to foreign aid in developing countries argued that the fact that investments needed to be complementary across sectors (to be successful, an iron ore mine might require a railroad, which in turn might require a port) meant that those investments needed to be centrally guided and planned.[14] William Easterly summarises this general view of the connections between development theory and policy:

> The poorest countries are in a *poverty trap* (they are poor *only* because they started poor) from which they cannot emerge

12 On a poverty trap see: Rosenstein-Rodan, Paul N. 1943. "Problems of Industrialisation of Eastern and South-Eastern Europe." *The Economic Journal* 53 (210/211); Sachs, Jeffrey. 2005. *The End of Poverty: How We Can Make It Happen in Our Lifetime* UK: Penguin Publishing.

13 The scale of development aid is significant. In 2017 official foreign aid was USD $146.6 billion. See: OECD. n.d. "Development Aid Stable in 2017 with More Sent to Poorest Countries."

14 The argument was that a marginal and piecemeal approach to investments might not create a sufficient rate of sustaining growth for 'take-off'. The social marginal products of investments might be below private marginal products due to complementarities, including pecuniary externalities (i.e. investments in one product might lower its price, benefiting another industry that uses it as an input). See, for instance: Murphy, Kevin M., Andrei Shleifer, and Robert W. Vishny. 1989. ":Industrialization and the Big Push." *The Journal of Political Economy* 97 (5); Gans, Joshua. 1998. "Industrialization Policy and the 'Big Push.'" In *Increasing Returns and Economic Analysis*, edited by Kenneth J. Arrow. Palgrave.

> without an aid-financed *Big Push*, involving investments and actions to address all constraints to development, after which they will have a *takeoff* into self-sustained growth, and aid will no longer be needed.[15]

Many free market economists have critiqued this diagnosis and treatment.[16] Calls for a "big push" of coordinated investments fail to explain how planners discover the information about the specific needs of citizens and what public goods will deliver those. Calls for foreign aid increases don't explain how reform can happen with failed governments and corrupt institutions. As the development economist Peter T. Bauer famously and controversially defined aid—"a transfer of resources from the *taxpayer* of a donor country to the *government* of a recipient country"—aid can lead to an enlargement of the public sector in developing economies.[17]

Simplistic one-size-fits-all analysis ignores the complexity of economies. It is not enough to copy, say, the constitution of the United States or the legal system of the United Kingdom and drop them into Haiti or Cambodia. Local policy conditions do matter. In the last few decades, the question of how to make poor countries rich has moved away from universal cure-alls towards more contextual experimentation

---

15 Emphasis in original. See: Easterly, William. 2006. *The White Man's Burden: Why the West's Efforts to Aid the Rest Have Done So Much Ill and So Little Good.* Penguin Publishing. p. 33.

16 Ibid.; Bauer, Peter T. 1976. *Dissent on Development.* US: Harvard University Press; Easterly, William. 2006. "Planners Versus Searchers in Foreign Aid." *Asian Development Review* 23 (2): pp. 1-35.

17 Emphasis added. See: Bauer, Peter T. 1975. "N H Stern on Substance and Method in Development Economics." *Journal of Development Economics* 2 (4). p. 396.

and policy innovation. Randomised control trials (RCTs), for instance, attempt to provide an empirical path to enlightenment on the causes of poverty and the implementation of solutions.[18] But while randomised evaluations can help specific causes of poverty, and to reveal specific solutions to those problems, they are still often followed by intervention and central planning that brushes over economic complexity, failing to tackle the underlying causes of poverty.

After the failure of the Soviet Union and the rise of China throughout the twentieth century, the key determinants of economic development have come to be understood as institutions and entrepreneurship.[19] Good institutions facilitate entrepreneurship and innovation which in turn drives long-run growth.[20] The prerequisites for economic prosperity in poor countries are the same as those in rich countries: prosperity requires high quality institutions that deliver economic, social and political freedom.[21] In many ways the very point of economic development is to secure individual liberty, as James Dorn writes:

---

18 See: Rodrik, Dani. 2008. "The New Development Economics: We Shall Experiment, but How Shall We Learn?" *Harvard Kennedy School Working Paper Series.*

19 Acemoglu, Daron, and James A. Robinson. 2012. *Why Nations Fail: The Origins of Power, Prosperity, and Poverty.* UK: Crown Publishers; North, Douglass C., John Joseph Wallis, and Barry R Weingast. 2009. *Violence and Social Orders: A Conceptual Framework for Interpreting Recorded Human History.* UK: Cambridge University Press; Weingast, Barry R. 1995. "The Economic Role of Political Institutions: Market-Preserving Federalism and Economic Development." *Journal of Law, Economics & Organization* 11 (1).

20 Carden, Art. 2009. "Can't Buy Me Growth: On Foreign Aid and Economic Change." *Journal of Private Enterprise* 25 (1).

21 Easterly, *The White Man's Burden: Why the West's Efforts to Aid the Rest Have Done So Much Ill and So Little Good*; Hausmann, Ricardo, and Dani Rodrik. 2003. "Economic Development as Self-Discovery." *Journal of Development Economics* 72 (2); Leeson and Boettke, "Two-Tiered Entrepreneurship and Economic Development."

> a spontaneous free-market process that enlarges individual choice is the true meaning of development; a coercive, state-led development policy that denies individuals the freedom to make their own choices is pseudo development.[22]

This begs the question: how can the developing world get better institutions? Historically there have been many ways to push institutional change. The most aggressive of which are militarised attempts to replace existing institutions—the belief that democracy could be imposed at the barrel of a gun that led to the invasion of Iraq in 2003. Institutional change in the pursuit of freedom, however, ought to be liberty-preserving. Institutions are particular, not generic: they need to be discovered by those who are most directly affected by them, and their implementation requires deep contextual information. Institutional reform is fundamentally an entrepreneurial problem. In the same way that business entrepreneurs look for opportunities and niches within a market, institutional entrepreneurs look for opportunities to create better rules.[23]

This is the role of institutional technologies: a generation of technological advances that allow entrepreneurs in the rich and poor worlds alike to experiment with new ways of coordinating economic exchange and social activity. Institutional technologies such as blockchains

---

22 Dorn, James A. 2002. "Economic Development and Freedom: The Legacy of Peter Bauer." *Cato Journal* 22 (2). p. 359.

23 Boettke, Peter J., Christopher J. Coyne, and Peter T. Leeson. 2008. "Institutional Stickiness and the New Development Economics." *American Journal of Economics and Sociology* 67 (2); Allen, Darcy W.E. 2019. "Entrepreneurial Exit: Developing the Cryptoeconomy." In *Blockchain Economics*, edited by Melanie Swan, et al. London, UK: World Scientific.

lower the costs of privately spinning up competing rule sets. From this perspective, blockchains aren't just another technology input into the economic development process. New privately-built institutional infrastructure can be developed without the support of extractive and corrupt states, or even in direct opposition to it—a process called institutional layering.

BLOCKCHAINS FOR PRIVATE ECONOMIC DEVELOPMENT

Given that many problems in the developing world derive from failures in central coordinating authorities, many observers and entrepreneurs have identified the profound implications of blockchains for development.[24]

For example, blockchain technology could be used to provide much more transparency in the ways rich countries support poor countries. As Mikayla Novak has pointed out, the fact that aid must go through intermediaries—whether those intermediaries are professional charitable organisations or foreign governments—before it can reach its intended recipients creates the risk that funds are siphoned or diverted against the wishes of the donor.[25] Blockchain technology could allow donors to track with high certainty exactly where their money goes.

Many of the tools of corrupt governments—such as land titles and

---

24 For instance, see: Pisa, Michael, and Matt Juden. 2017. "Blockchain and Economic Development: Hype Vs. Reality." *Center for Global Development Policy,* Paper No. 107; Adams, Richard, Beth Kewell, and Glenn Parry. 2018. "Blockchain for Good? Digital Ledger Technology and Sustainable Development Goals." In *Handbook of Sustainability and Social Science Research*, edited by Walter Leal Filho, Robert W Marans, and John Callewaert. Springer.

25 Novak, Mikayla. 2018. "Crypto-Altruism: Some Institutional Economic Considerations." *SSRN 3230541.*

other registries—can be made more transparent and decentralised.[26] Blockchains can automate many aspects of consensus and verification over shared facts, such as identity, ownership and transactions in the absence of a central coordinating administrative authority.[27] Blockchains can free people from the shackles of state-maintained identities through self-sovereign identity that can be verified without relying on the state. These technologies also facilitate movement across borders, and for displaced people, by establishing identity that can easily be verified without relying on a central authority. Licensing could also be taken away from bureaucratic and technocratic central governments and replaced with private, competing, verifiable alternatives.[28]

But adopting such a technology relies on a certain amount of trust in existing institutions to implement that change. Property registries, for instance, are a clear and obvious blockchain use-case. Existing national

---

26 See, for instance: McMurren, Juliet, Andrew Young, and Stefaan Verhulst. 2019. "Addressing Transaction Costs through Blockchain and Identity in Swedish Land Transfers." *GovLab*; Thakur, Vinay et al. 2019. "Land Records on Blockchain for Implementation of Land Titling in India." *International Journal of Information Management*.

27 Berg, Davidson, and Potts, *Understanding the Blockchain Economy: An Introduction to Institutional Cryptoeconomics*; Davidson, Sinclair, Primavera De Filippi, and Jason Potts. 2018. "Blockchains and the Economic Institutions of Capitalism." *Journal of Institutional Economics* 13 (4); Catalini, Christian and Joshua S. Gans. 2016. "Some Simple Economics of the Blockchain." *Rotman School of Management Working Paper No. 2874598*; *MIT Sloan Research Paper No. 5191-16*.

28 Adams, Richard, Beth Kewell, and Glenn Parry. 2017. "Blockchain for Good? Digital Ledger Technology and Sustainable Development Goals." In *Handbook of Sustainability and Social Science Research*. Pp. 127-140; Allen, Darcy W.E., and Chris Berg. 2018. "Blockchain for Development." Melbourne, Australia: RMIT APEC Study Centre; GSMA. 2017. "Blockchain for Development." UK: GSMA; Pisa and Juden, "Blockchain and Economic Development: Hype Vs. Reality."

or jurisdictional titling systems could theoretically be transitioned to a distributed or decentralised blockchain-based titling system. While Sweden has been investigating putting its property titles on blockchain since 2016, it already has one of the world's most famously efficient governments. The value for Sweden for property titles doesn't seem transformative. By contrast think of a developing economy, such as Haiti, trying to transition from centralised to decentralised property titles. Blockchain infrastructure in Haiti could help to unlock the tens of billions of dollars in dead capital locked up by inadequate and insecure property rights.[29] But this transition requires an act of trust in central governments in the first place.[30]

Using blockchains in developing economies must tackle this paradox of trust. Any proposal for public policy reform must account for the institutional problems that have created bad policy in the first place.[31]

The problem here is that institutions tend to be monopolies. One country has one court system, one bureaucracy in charge of property titles, one authority giving out birth certificates. To get better institutions, we have to replace the corrupt old ones, and that's hard to do, especially given the intransigence of rent-seekers who benefit from them.[32] Blockchain enables citizens to break this relationship between

---

29 On dead capital see: De Soto, Hernando. 2000. *The Mystery of Capital: Why Capitalism Triumphs in the West and Fails Everywhere Else.* Basic Civitas Books.

30 Graglia, J. Michael, and Christopher Mellon. 2018. "Blockchain and Property in 2018: At the End of the Beginning." *Innovations: Technology, Governance, Globalization* 12 (1-2).

31 On this point see: Boettke, Coyne, and Leeson, "Institutional Stickiness and the New Development Economics."

32 Allen, Darcy W.E., and Chris Berg. 2019. "Why Blockchain Could Be the Key to Solving the World's Development Problems." *Foundation for Economic Education.*

geographical area and institutional rules—the technology offers a mechanism of non-territorial exit, or cryptosecession.[33] Rather than having to physically exit the jurisdiction of the nation state, people can live within the same geographical borders but non-territorially exit to blockchain-based institutions.

Critically, this institutional discovery process can occur without displacing existing state-based geographical institutions. As some scholars of institutional change have pointed out, rarely is the experience of reform a sharp break between new institutions and old institutions. The scholarship around institutional layering describes how the practice of reform involves bolting institutions together, supplementing, complementing and deprecating the old with the new.[34] Except in a few circumstances—where catastrophe or invasion shocks the system into wholesale change—reform is almost always incremental, not revolutionary.

The process of institutional layering is messy. It doesn't fit easily within a grand plan that development professionals or academics may desire. It involves a process of competing institutional systems. Blockchains mean that institutional change is no longer restricted to change across entire political jurisdictions. We can now have low-cost competing mixes of private institutions layering on top of existing ones. There are several potential implications for this understanding, not least the potential for more permissionless institutional change, polycentric

---

33 MacDonald, *The Political Economy of Non-Territorial Exit*, 63. On cryptosecession also see: MacDonald, Trent. 2015. "Theory of Non-Territorial Internal Exit." *SSRN 2661226*; MacDonald, Allen, and Potts, "Blockchains and the Boundaries of Self-Organized Economies: Predictions for the Future of Banking."

34 Heijden, Jeroen Van der. 2011. "Institutional Layering: A Review of the Use of the Concept." *Politics* 31 (1).

private governance and more voluntary choice between rulesets.

One compelling opportunity for institutional layering through blockchain is the legal system. In a well-functioning economy, judges act as a neutral arbitrator between parties.[35] But many developing countries suffer from poorly functioning judicial systems. When judges are susceptible to bribery, extortion, or intimidation then the consequences for business investment, entrepreneurial activity and ultimately economic growth can be severe.[36] The phrase "why buy a lawyer when you can buy a judge" captures how a legal system can be debased by corruption.[37]

As we saw in Chapter 3, promises encoded in smart contracts are much harder to terminate than promises made in traditional business agreements.[38] From a macro-institutional lens, to the extent that contractual relationships can be specified in code, smart contracts provide a mechanism for contracting parties to avoid corrupt legal systems. Each party can then be satisfied that exchanges of value will be automatically triggered when contract terms are met.

Blockchain-based legal systems can co-exist with existing legal

---

35 Djankov, Simeon et al. 2003. "The New Comparative Economics," *Journal of Comparative Economics* 31 (4); Shleifer, Andrei. 2005. "Understanding Regulation." *European Financial Management* 11 (4).

36 Acemoglu, Daron, and James A. Robinson. 2010. "The Role of Institutions in Growth and Development." In *Leadership and Growth*, edited by David Brady and Michael Spence. Washington, DC: The World Bank; Mauro, Paolo. 1995. "Corruption and Growth." *The Quarterly Journal of Economics* 110 (3); Shleifer, Andrei, and Robert W. Vishny. 1993. "Corruption."

37 United Nations. 2005. "Report of the Fourth Meeting of the Judicial Integrity Group." Vienna, Austria: United Nations Office on Drugs and Crime. At p. 6.

38 Wright and De Filippi, *Blockchain and the Law: The Rule of Code*.

systems and don't require wholesale reform.[39] Polycentric legal systems already exist. Private law co-exists with public law, and has done so at least since the development of commercial law in the Islamic and Christian middle ages.[40] Commercial dispute resolution services provide legal systems for contractual disputes that are subject to competitive dynamics. Contracting parties agree to use a commercial arbitration service that each sees as unlikely to favour the other party, and this places pressure on those services to compete on that margin. An easily corruptible commercial dispute resolution service will not last long. The rise of alternative arbitration services also places pressure on public legal systems.[41] International commercial courts play an increasingly important role in the global market, as multinational firms seek to do business between countries with significantly different legal

---

39 Berg, Chris, Sinclair Davidson, and Jason Potts. 2019. "Towards Crypto-Friendly Public Policy." In *Blockchain Economics: Implications for Distributed Ledgers*, edited by Melanie Swan, et al. London, UK: World Scientific; Novak, Mikayla. (forthcoming) "Crypto-Friendliness: Understanding Blockchain Public Policy." *Journal of Entrepreneurship and Public Policy*.

40 On development of Islamic commercial law see: Greif, Avner. 2006. *Institutions and the Path to the Modern Economy: Lessons from Medieval Trade.* UK: Cambridge University Press; Greif, Avner. 1993. "Contract Enforceability and Economic Institutions in Early Trade: The Maghribi Traders' Coalition." *The American Economic Review* 83 (3). On Christian developments see: Berman, *Law and Revolution*; Milgrom, Paul R., Douglass C. North, and Barry R. Weingast. 1990. "The Role of Institutions in the Revival of Trade: The Law Merchant, Private Judges, and the Champagne Fairs." *Economics & Politics* 2 (1); Trakman, *The Law Merchant: The Evolution of Commercial Law*.

41 Dakolias, Maria. 1995. "A Strategy for Judicial Reform: The Experience in Latin America." *Virginia Journal of International Law* 36; Moyer, Thomas J., and Emily Stewart Haynes. 2002. "Mediation as a Catalyst for Judicial Reform in Latin America." *Ohio State Journal on Dispute Resolution* 18 (3).

systems, traditions, and institutional quality.[42]

They provide freedom to citizens of developing economies to choose their jurisdiction and to facilitate exchange without being held by existing corrupt legal systems. Smart contracts are the foundation for greater technological freedom in the developing world. This freedom opportunity will only expand as the tokenisation of assets increases and financial value is more widely represented on distributed ledgers. We're about to enter a world where corrupt legal systems have competitors—decentralised, censorship-resistant competitors.

Institutional layering is fundamentally liberty-friendly. Institutional entrepreneurs propose new institutions and citizens voluntarily select between alternatives. This process contrasts starkly with top-down change that affects entire jurisdictions. But more fundamentally, the form of institutional layering we have described here doesn't require permission from government. It doesn't require political will or consent to catalyse change. Entrepreneurs can offer alternative dispute resolution services through smart contracts without ever having to ask government for permission to do so. The underlying institutional framework—the real-world legal system, or financial system, or bureaucratic organisation—remains untouched.

In that sense, decentralised blockchain development aligns more closely with the concept of permissionless innovation—outlined by Adam Thierer—which is generally used to describe innovation on the internet, rather than the economic development of the global south.[43]

---

42 Middleton, John. 2018. "The Rise of the International Commercial Court." In *The 2018 Hong Kong International Commercial Law Conference, Hong Kong.*

43 Thierer, Adam D. 2014. *Permissionless Innovation: The Continuing Case for Comprehensive Technological Freedom.* Mercatus Center at George Mason University.

Budding entrepreneurs within developing economies—or private external people—can introduce competing rule sets to solve social, economic and political coordination problems. They seek to gain users on their platforms and can act as the residual claimants on the profits of doing so.

Real world institutions are difficult to change partly because of the corruption and rent seeking that entrenches them. Corrupt judges or bureaucrats, for example, have come to rely on the proceeds of corruption and will fight to defend them. Iterative institutional change involves displacing the rent seeking that has built up. Poor countries face significant political and economic costs of transitioning from legacy governance.[44] In this way institutional layering could help to ameliorate institutional path dependency that plagues developing economies.

Similarly, blockchain based institutional laying is not susceptible to the fallacy of planning. Rather than a central planner, foreign aid bureaucrat, or development economist determining what institutional reforms are required—or even precisely what the economic development problem is—in a process of decentralised blockchain development individual entrepreneurs are incentivised and enabled to create new institutions. Citizens are free to use the new institutions only if they are desirable.

For example, institutional entrepreneurs might set up competing systems of property rights protection and enforcement, and over time groups of people might come to consensus that the blockchain-based system is the true and correct state. In this way entrepreneurs can use blockchain to create a new de facto system of property titling. It would

44 Tullock, Gordon. 1975. "The Transitional Gains Trap." *The Bell Journal of Economics* 6 (2).

then potentially be up to the existing courts system (or indeed some new blockchain-based dispute resolution system) to resolve disputes.

Farsighted governments could welcome this sort of innovation, allowing disputes that might have arisen in the digital world to be resolved using terrestrial courts. Alternatively, in countries where the government is hostile to innovation, blockchain entrepreneurship could take the form of evasive entrepreneurship.[45] Either path, however, is comparatively more permissionless than a centralised process of institutional reform, because that would require collective action solutions to the dissipation of any political or economic rents that have built up.

The process of institutional change that blockchain facilitates should generate longer lasting and more widely adopted institutions—that is, institutions that are *sticky*. Governance arrangements that are more institutionally sticky align with spontaneously emergent (or 'indigenous') institutions (including informal norms).[46] The more decentralised and competitive the entrepreneurial economic development process is, the stickier we would expect it to be. A more decentralised development process draws more closely on local knowledge from multiple sources. Institutions that emerge through the distributed action of entrepreneurs—rather than those that are planned by top-down planners—"tend to be the stickiest institutions of all."[47]

Blockchain is a technology of community consensus, not coercion or political authority. From a computer science perspective it represents

---

45 Elert, Niklas, Magnus Henrekson, and Joakim Wernberg. 2016. "Two Sides to the Evasion: The Pirate Bay and the Interdependencies of Evasive Entrepreneurship." *Journal of Entrepreneurship and Public Policy* 5 (2).

46 Boettke, Coyne, and Leeson, "Institutional Stickiness and the New Development Economics."

47 Ibid., p. 339.

a new architecture of trust.[48] But from a social or cultural perspective, it resembles the institutions of many traditional societies where communities come to agreement about shared facts as a group. For example, many people in the poorest countries lack the formal identification necessary to engage fully within the local or regional economy. It is not that they lack technologies of identity, but that their identity 'management' involves networks of community consensus about who owns what, who is related to who, and who has authority to make decisions. Local communities often resist attempts by the national government to impose formal identities on them, concerned about the risk that with formal identity comes repression.[49] These new technologies offer the possibility of new identity institutions that are developed by a community, maintained by the consensus of that community, but are simultaneously able to scale and be treated as authoritative by those outside the community.

At the same time the new competitive dynamic between the existing public institutions and the new self-governing institutions might create pressures to improve existing ones, as rent-seekers discover that citizens and consumers prefer to use the self-governing ones. We also expect permissionless institutional layering to be more *adaptive* through time as local conditions change. That is, we not only anticipate the initial entrepreneurial action to better match local conditions and problems, but that this process to better deal with adaptation when the conditions necessarily change.

---

48 Werbach, Kevin. 2018. *The Blockchain and the New Architecture of Trust.* Cambridge, MA: MIT Press.

49 Berg, Chris, Sinclair Davidson, and Jason Potts. 2017. "The Institutional Economics of Identity." *SSRN Paper No. 3072823*; Berg, Alastair et al., 2018. "Identity as Input to Exchange." *SSRN Paper No. 3171960.*

Blockchain and other distributed ledger technologies might better connect local contextual knowledge with the allocation of entrepreneurial investments and institutions, as an alternative solution to the problem the "big push" is seeking to ameliorate. The second part of the "big push" argument is that societies need many complementary investments across an economic system in order to escape a poverty trap and to "take-off." That is, greater visibility of trading platforms—through the provision of new infrastructure for trade and supply chains—gives entrepreneurs better visibility of potential market opportunities.[50]

Many theories of economic development were well designed for political intervention. A savings-investment gap could be closed by concerted effort in the form of foreign aid and other interventions. A lack of coordinated complementary investments could be remedied through a "big push" of state-directed investment. There has been a range of critiques to these understandings, pushing analysis back to unreliable institutions and impediments to entrepreneurial activity.

But restructuring institutions has turned out to be hard. Corrupt and inefficient institutions as well as regulatory hurdles that protect incumbent firms from competition have constituencies that enjoy rent from the status quo. Free market economists have rarely been able to offer concrete solutions to poverty—it is little help to implore poor countries just to "be less corrupt."

That has now changed. Institutional technologies such as blockchains allow for institutional entrepreneurship without institutional destruction. Blockchains can co-exist with the existing legal and regulatory institutions that do the same. Entrepreneurs can now spin-up new

---

50 Kirzner, Israel M. 1997. "Entrepreneurial Discovery and the Competitive Market Process: An Austrian Approach." *Journal of Economic Literature* 35 (1).

institutional forms without needing to make bargains with (or alienate) the beneficiaries of incumbent institutions. In countries that suffer from weak, corrupt or otherwise ineffective institutions, technology gives entrepreneurs the ability to layer new institutions on top of old, without the need for the permission of incumbent beneficiaries, and offers a promising new approach to development.

09.

CONCLUSION

We need a new strategy for expanding our individual freedoms in the twenty-first century. Today a remarkable suite of technologies—from blockchains to machine learning—can be used to escape, undermine and outcompete nation states. We should be deeply optimistic about scalable private currencies, distributed legal systems and censorship-resistant organisations. These are technologies of freedom. But these opportunities for liberty must be built. The role of the liberty movement must shift—away from being passive consumers or defenders of technological progress, towards active entrepreneurship, tinkering and adopting these new technologies.

The free market movement needs to shift course from activism through politics—of asking the state for freedom—towards entrepreneurial discovery. Individual freedom over coming decades won't come through voting at the ballot box, but rather through the spread of online, distributed and privacy-enhancing technologies. These technologies will be the foundation for new types of voluntary social organisation, the next phase of globalisation and a new era of individual autonomy.

For decades the liberty movement has sought change through the machinations of politics. This is an attempt to use the very apparatus that we claim threatens those liberties—the state—to take back our liberties. We have morally and rationally convinced students, peers and politicians about what reforms are desirable. Think tanks research issues and run campaigns. Scholars have written books and articles. We have studied the great liberal thinkers throughout history. And to be sure this approach has had its successes—there have been periods of reform and retractions of state power. But in a world of low-cost digital technologies, this strategy is no longer optimal.

Blockchains, smart contracts, automation and encryption are tools for a new wave of digital experimentation and freedom. This new type of activism that we have proposed—and the bottom-up wave of social and economic change that comes from it—is fundamentally entrepreneurial. This is a step-change in how we expand our freedoms. We don't beg the state: we make state power redundant, ineffective and obsolete by building better private institutions.

One of our aims in this book was to push back on techno-pessimism. Many of the technologies we have described conjure popular images of totalitarian surveillance states. Our entrepreneurial vision is deeply optimistic about these technologies being tools for freedom. The solution to threats over our freedom is more freedom. It is up to us to apply technologies in ways that allow us and others to escape and evade repressive states. It's no longer enough to talk about reform; it's time to contribute to a world where people can choose to organise their lives using technologies rather than governments. In practice that means learning about frontier technological advances, downloading privacy-enhancing messaging applications, and developing new economic institutions that have freedom built in.

While our proposition may seem quite radical, in many ways it is a manifestation of many familiar ideas for the liberty-minded: a process of private, open and voluntary change that is pushed by entrepreneurs and chosen by autonomous individuals. The strategy we have described embraces the principles of freedom. The way that we organise our economic, social and political lives soon won't be dictated by territorial monopoly or path dependency. They will emerge from the choices of the entrepreneurs building platforms and the consumers deciding to use them. We can now have those same principles for freedom—we are about to enter a world where we have a much wider choice over what rules we live under, the privacy protections we take, and the way we exchange and share with others. New technologies should be embraced with the understanding that they are kept open, that they are competitive, and they protect our personal and property rights. Technologies augment our freedoms—it's our job to apply them. This process will be messy, competitive and evolutionary.

There are several other benefits to adopting the strategy and vision we've set out here. The pursuit of freedom becomes global. We are no longer tied to reform within borders or jurisdictions or bound by a political election cycle. This also implies that the opportunity for impact for liberty-enhancing entrepreneurship goes far beyond the jurisdiction within which it was developed. What's more, the stakes are high because these global digital platforms—enforcing property rights and enabling the transfer of value—are being adopted today.

The technological developments we have outlined in this book have the potential to empower individual choice and undermine major areas of government. Our selection is only a tiny subset of the technological advances on the horizon. We are seeing rapid advances in technologies for bio-enhancement through genomics, additive manufacturing, internet of things including proof of location services, drones and robotics, augmented reality, and many more.

Technological advances are highly complementary products and services. A cryptocurrency combined with self-sovereign identity networked into a distributed autonomous organisation is more powerful than any of these technologies on their own. Furthermore, the nature of digital platforms, including their network effects, mean that they need to be bootstrapped. This makes it imperative that whether you are a technology entrepreneur seeking to expand individual freedom, or simply want to spread the ideas of freedom—now is the time to build the technologies of freedom, share those ideas and platforms with others, and become active participants in a freer and more prosperous digital society.

REFERENCES

ABC News Australia. 2019. "Seasteading Bitcoin Couple Charged with Violating Thai Sovereignty as Navy Boards Floating Home." *ABC News Australia*, April 21.

Abelson, Hal et al. 1997. "The Risks of Key Recovery, Key Escrow, and Trusted Third-Party Encryption." *World Wide Web Journal* 2 (3): pp. 241-257.

ACCC. 2019. "Australian Competition and Consumer Commission Digital Platforms Inquiry Final Report." *ACCC.*

Acemoglu, Daron, and James A. Robinson. 2012. *Why Nations Fail: The Origins of Power, Prosperity, and Poverty.* UK: Crown Publishers.

———. 2010. "The Role of Institutions in Growth and Development." In *Leadership and Growth*, edited by David Brady and Michael Spence. Washington, DC: The World Bank.

———. 2000. "Political Losers as a Barrier to Economic Development." *American Economic Review* 90 (2): pp. 126-130.

Adams, Douglas. *The Restaurant at the End of the Universe*. Pan Books, 1980.

Adams, Richard, Beth Kewell, and Glenn Parry. 2018. "Blockchain for Good? Digital Ledger Technology and Sustainable Development Goals." In *Handbook of Sustainability and Social Science Research*, edited by Walter Leal Filho, Robert W. Marans, and John Callewaert. Springer.

Agrawal, Ajay, Joshua S. Gans, and Avi Goldfarb. 2018. *Prediction Machines: The Simple Economics of Artificial Intelligence* Harvard Business Review Press.

Aggarwal, Aradhna. 2007. "Impact of Special Economic Zones on Employment, Poverty and Human Development." *Indian Council for Research on International Economic Relations (ICRIER) Working Paper No. 194*.

AidCoin. n.d. "AidCoin Whitepaper." https://www.aidcoin.co/assets/documents/whitepaper.pdf?v=1.0.2.

Aït-Kacimi, Nessim, and Raphaël Bloch. 2019. "The Banque De France Will Experiment with a Digital Euro in 2020." *Les Echos*, December 4.

Akinci, Gokhan, and James Crittle. 2008. "Special Economic Zones: Performance, Lessons Learned, and Implications for Zone Development." Washington, DC: The World Bank Group.

Alchian, Armen A., and Harold Demsetz. 1972. "Production, Information Costs, and Economic Organization." *The American Economic Review* 62 (5).

Al Jazeera. 2019. "Paris, Rome, Berlin Preparing to Block Facebook's Libra in Europe." *Al Jazeera*, October 18.

Allcott, Hunt, and Matthew Gentzkow. 2017. "Social Media and Fake News in the 2016 Election." *Journal of Economic Perspectives* 31 (2): 211-236.

Allen, Darcy W.E. 2019. "Entrepreneurial Exit: Developing the Cryptoeconomy." In *Blockchain Economics*, edited by Melanie Swan, et al. London, UK: World Scientific.

Allen, Darcy W E, Dick Auer, Chris Berg, Gus Hurwitz, Aaron M Lane, Geoffrey A Manne, Julian Morris, and Jason Potts. 2019. "Submission on the Final Report of the Australian Competition and Consumer Commission's Digital Platforms Inquiry." *Chris Berg*, September 12.

Allen, Darcy W.E., Alastair Berg, Chris Berg, Brendan Markey-Towler, and Jason Potts. 2019. "Some Economic Consequences of the GDPR." *Economics Bulletin* 39 (2): pp. 785-797.

Allen, Darcy W.E., Alastair Berg, and Brendan Markey-Towler. 2019. "Blockchain and Supply Chains: V-Form Organisations, Value Redistributions, De-Commoditisation and Quality Proxies." *The Journal of the British Blockchain Association* 2 (1): pp. 1-8.

Allen, Darcy W.E., and Chris Berg. 2019. "Why Blockchain Could Be the Key to Solving the World's Development Problems." *Foundation for Economic Education.*

———. 2018. *Australia's Red Tape Crisis: The Causes and Costs of over-Regulation*. Brisbane, Australia: Connor Court Books.

———. 2018. "Blockchain for Development." Melbourne, Australia: RMIT APEC Study Centre.

———. 2017. "Subjective Political Economy." *New Perspectives on Political Economy* 13 (1-2): pp. 19-40.

———. 2014. "The Sharing Economy: How over-Regulation Could Destroy an Economic Revolution." *Institute of Public Affairs*, December.

Allen, Darcy W E, Chris Berg, Sinclair Davidson, Mikayla Novak, and Jason Potts. 2019. "International Policy Coordination for Blockchain Supply Chains." *Asia & the Pacific Policy Studies* 6 (3).

Allen, Darcy W.E., and Aaron M. Lane. 2020. "Cryptodemocratic Governance in Special Economic Zones." *Journal of Special Jurisdictions* 1 (1).

Allen, Darcy W.E., Aaron M. Lane, and Marta Poblet. 2019. "The Governance of Blockchain Dispute Resolution." *Harvard Negotiation Law Review* 25: pp. 75-101.

Allen, Darcy W.E., Chris Berg, Aaron M. Lane, and Jason Potts. 2018. "Cryptodemocracy and Its Institutional Possibilities." *The Review of Austrian Economics*: pp. 1-12.

Allen, Darcy W.E., Chris Berg, and Aaron M. Lane. 2019. *Cryptodemocracy: How Blockchain Can Radically Expand Democratic Choice*, edited by Lenore T. Ealy and Paul Aligica. Lanham, US: Lexington Books.

Allen, Darcy W.E., Chris Berg, and Jason Potts. 2019. "Blockchain Technology and the Theory of Economic Development." *SSRN Paper No. 3333568.*

Arendt, Hannah. 1972. *Crises of the Republic: Lying in Politics, Civil Disobedience on Violence, Thoughts on Politics, and Revolution.* Harcourt Brace Jovanovich.

Aridor, Guy et al. 2020. "The Economic Consequences of Data Privacy Regulation: Empirical Evidence from GDPR." *SSRN 3522845.*

Arrow, Kenneth J. 1962. "Economic Welfare and the Allocation of Resources for Invention." In *The Rate and Direction of Inventive Activity: Economic and Social Factors*, edited by Richard R. Nelson. Princeton, New Jersey: Princeton University Press.

Asimov, Isaac. 1986. *Robots and Empire.* HarperCollins.

———. 1942. "Runaround." *Astounding Science-Fiction* 29 (1).

Atlas Network. 2019. "The Templeton Freedom Award."

Attwood, Lynne. 2017. *Gender and Housing in Soviet Russia: Private Life in a Public Space.* Manchester, UK: Manchester University Press.

Back, Adam. 1997. "A Partial Hash Collision Based Postage Scheme." *Hashcash.org*, March 28.

Bagdikian, Ben H. 2004. *The New Media Monopoly.* Penguin Random House.

Banjo, Shelly, and Lulu Yilun Chen. 2019. "Digital Dissidents Are Fighting China's Censorship Machine." *Bloomberg Businessweek*, June 4.

Barbrook, Richard, and Andy Cameron. 1995. "The Californian Ideology." *Mute*, September 1.

Bauer, Peter T. 1976. *Dissent on Development.* US: Harvard University Press.

———. 1975. "N H Stern on Substance and Method in Development Economics." *Journal of Development Economics* 2 (4).

Bell, Stephen, and Andrew Hindmoor. 2009. *Rethinking Governance: The Centrality of the State in Modern Society.* Cambridge, UK: Cambridge University Press.

Bell, Tom W. 2017. *Your Next Government? From the Nation State to Stateless Nations*. Cambridge: Cambridge University Press.

Benos, Evangelos, Rod Garratt, and Pedro Gurrola-Perez. 2017. "The Economics of Distributed Ledger Technology for Securities Settlement." (Available at SSRN: https://ssrn.com/abstract=30237792017).

Berg, Alastair et al., 2018. "Identity as Input to Exchange." *SSRN Paper No. 3171960.*

Berg, Chris. 2018. *The Classical Liberal Case for Privacy in a World of Surveillance and Technological Change.* Palgrave Macmillan.

———. 2018. "Regulation and Red Tape in a Small Open Economy." In *Australia's Red Tape Crisis: The Causes and Costs of Over-Regulation*, ed. Darcy WE Allen and Chris Berg. Queensland: Connor Court Publishing.

———. 2016. "Safety and Soundness: An Economic History of Prudential Bank Regulation in Australia, 1893-2008." *Thesis submitted to RMIT University.*

———. 2012. *In Defence of Freedom of Speech: From Ancient Greece to Andrew Bolt.* Monographs on Western Civilisation. Melbourne, Australia: Institute of Public Affairs.

———. 2008. *The Growth of Australia's Regulatory State: Ideology, Accountability and the Mega-Regulators*. Melbourne, Australia: Institute of Public Affairs.

Berg, Chris, Sinclair Davidson, and Jason Potts. 2020. "Proof of Work as a Three-Sided Market." *Frontiers in Blockchain*, January 31.

———. 2020. "Capitalism after Satoshi: Blockchains, dehierarchicalisation, innovation policy, and the regulatory state." *Journal of Entrepreneurship and Public Policy* 9 (2): pp. 152-164.

———. 2019. *Understanding the Blockchain Economy: An Introduction to Institutional Cryptoeconomics*. Cheltenham, UK: Edward Elgar Publishing.

———. 2019. "Towards Crypto-Friendly Public Policy." In *Blockchain Economics: Implications for Distributed Ledgers*, edited by Melanie Swan, et al. London, UK: World Scientific.

———. 2018. "Outsourcing Vertical Integration: Distributed Ledgers and the V-Form Organisation." *SSRN* Paper no. 3300506.

———. 2017. "The Institutional Economics of Identity." *SSRN Paper No. 3072823.*

Berman, Ana. 2019. "Paypal Originally Aimed to Create Global Currency Similar to Crypto, Co-Founder Admits." *Cointelegraph*, February 1.

Berman, Harold Joseph. 1983. *Law and Revolution.* Cambridge, Massachusetts: Harvard University Press.

Bernanke, Ben. 2002. "On Milton Friedman's Ninetieth Birthday." *Conference to Honor Milton Friedman, University of Chicago, Chicago, Illinois.*

Bernanke, Naomi. 2007. *The Shock Doctrine: The Rise of Disaster Capitalism*. Toronto, Canada: Alfred A. Knopf.

Bernstein, William J. 2009. *A Splendid Exchange: How Trade Shaped the World.* Grove Press.

Bestagini, Paolo, et al. 2013. "Local Tampering Detection in Video Sequences." *Paper presented at the 2013 IEEE 15th International Workshop on Multimedia Signal Processing (MMSP), September 30 - October 2.*

Bitouk, Dmitri, Neeraj Kumar, Samreen Dhillon, Peter Belhumeur, and Shree K. Nayarl. 2008. "Face Swapping: Automatically Replacing Faces in Photographs." *ACM Transactions on Graphics (TOG)* 27 (3).

Boettke, Peter J. 2019. "The Role of the Economist in a Free Society." *Coordination Problem*, August 7.

———. 2019. "The Role of the Economist in a Free Society." *Address given at the Mont Pelerin Society regional meeting in Dallas,*

Texas, May 17-19, 2019.

———. 2018. *F. A. Hayek: Economics, Political Economy and Social Philosophy*. London, UK: Palgrave MacMillan.

———. 2012. *Living Economics: Yesterday, Today, and Tomorrow.* Independent Institute.

———. 2007. "Liberty Vs. Power in Economic Policy in the 20th and 21st Centuries." *The Journal of Private Enterprise* 22 (2): pp. 7-36.

Boettke, Peter J., and Matthew D. Mitchell. 2017. *Applied Mainline Economics: Bridging the Gap between Theory and Public Policy*. Arlington, Virginia: Mercatus Center at George Mason University.

Boettke, Peter J., S. Haeffele-Balch, and V.H. Storr. 2016. *Mainline Economics: Six Nobel Lectures in the Tradition of Adam Smith*. Mercatus Center, George Mason University.

Boettke, Peter J., Christopher J. Coyne, and Peter T. Leeson. 2008. "Institutional Stickiness and the New Development Economics." *American Journal of Economics and Sociology* 67 (2).

Bohme, Rainer, and M Kirchner. 2013. "Counter-Forensics: Attacking Image Forensics." Chapter 12 in *Digital Image Forensics: There Is More to a Picture Than Meets the Eye*, edited by Husrev Taha Sencar and Nasir Memon. New York: Springer-Verlag.

Bostrom, Nick. 2014. *Superintelligence: Paths, Dangers, Strategies.* Oxford, UK: Oxford University Press.

Bradsher, Keith. 2012. "Market's Echo of Tiananmen Date Sets Off Censors." *The New York Times*, June 4.

Breland, Ali. 2019. "The Bizarre and Terrifying Case of the "Deepfake" Video That Helped Bring an African Nation to the Brink."

Mother Jones, March 15.

Brennan, Jason. 2016. *Against Democracy*. New Jersey, US: Princeton University Press.

Bringsjord, Selmer, Paul Bello, and David Ferrucci. 2003. "Creativity, the Turing Test, and the (Better) Lovelace Test." In *The Turing Test*. Springer. Pp. 215-239.

British Security Industry Association. 2013. "The Picture Is Not Clear: How Many CCTV Surveillance Cameras Are There in the UK?" *British Security Industry Association.*

Brunner, Karl, and Allan H. Meltzer. 1971. "The Uses of Money: Money in the Theory of an Exchange Economy." *American Economic Review* 61 (5): 784-805.

Buchanan, James M. 2014. *Fiscal Theory and Political Economy: Selected Essays*. Chapel Hill, North Carolina: UNC Press Books.

Buchanan, James M., and Gordon Tullock. 1962. *The Calculus of Consent*, vol. 3. Ann Arbor, MI: University of Michigan Press.

Buchanan, James M., and Viktor J. Vanberg. 1991. "The Market as a Creative Process." *Economics and Philosophy* 7 (2): pp. 167-186.

Burrell, Jenna. 2016. "How the Machine 'Thinks': Understanding Opacity in Machine Learning Algorithms." *Big Data & Society* 3 (1): 1-12.

Buterin, Vitalik. 2017. "Change the Incentives, Change the World." *Cato Unbound*, June 16.

———. 2014. "I'm Not Understanding Why Dominant Assurance Contracts Are So Special." *Ethereum Community Forum.*

Bylund, Per L., and Matthew McCaffrey. 2017. "A Theory of

Entrepreneurship and Institutional Uncertainty." *Journal of Business Venturing* 32 (5).

Cadell, Cate, and Pei Li. 2017. "Tea and Tiananmen: Inside China's New Censorship Machine." *Reuters*, September 29.

Cadwalladr, Carole. 2019. "Facebook's Role in Brexit – and the Threat to Democracy." *TED Talk*, April.

Cai, Yuanfeng, and Dan Zhu. 2016. "Fraud Detections for Online Businesses: A Perspective from Blockchain Technology." *Financial Innovation* 2 (20).

Caplan, Bryan. 2011. *The Myth of the Rational Voter: Why Democracies Choose Bad Policies*. New Jersey, US: Princeton University Press.

Carden, Art. 2009. "Can't Buy Me Growth: On Foreign Aid and Economic Change." *Journal of Private Enterprise* 25 (1).

Casey, Michael, et al. 2018. *The Impact of Blockchain Technology on Finance: A Catalyst for Change.* International Center for Monetary and Banking Studies (ICBM).

Catalini, Christian and Joshua S. Gans. 2016. "Some Simple Economics of the Blockchain." *Rotman School of Management Working Paper No. 2874598*; *MIT Sloan Research Paper No. 5191-16.*

Catchlove, Paul. 2017. "Smart Contracts: A New Era of Contract Use." *SSRN 3090226.*

Cerviere, Giacinto. 2009. "The Rose Island." *Abitare.*

Chamberlain, Will. 2019. "Platform Access Is a Civil Right." *Human Events*, May 3.

Chaum, David. 1985. "Security without Identification: Transaction Systems to Make Big Brother Obsolete." *Communications of the ACM* 28 (10).

Chong, Nick. 2019. "European Central Bank Wants to Beat Other Countries in Digital Currency Game." *Blockonomi*, December 14.

Clark, Joseph. 2006. "Shares in People." *Policy* 22 (1): 3.

Clinton, Hillary. 2010. "Remarks on Internet Freedom." *Exhibit in The Newseum*, January 21.

Cochran, Matthew. 2015. "Conservatism Is Obsolete." *The Federalist*, March 9.

Cole, Samantha. 2017. "AI-Assisted Fake Porn Is Here and We're All Fucked." *Vice*, December 12.

Coyne, Christopher J. and Peter T. Leeson. 2004. "The Plight of Underdeveloped Countries." *Cato Journal* 24 (3): 235-249,

Crane, George T. 1994. "'Special Things in Special Ways': National Economic Identity and China's Special Economic Zones." *The Australian Journal of Chinese Affairs* (32).

———. 1990. *The Political Economy of China's Special Economic Zones*. Routledge.

Crawford, Michael. 1970. "Money and Exchange in the Roman World." *The Journal of Roman Studies* 60 (1): 40-48.

Crews, Clyde Wayne. 2019. "Ten Thousand Commandments 2019." Washington DC: Competitive Enterprise Institute.

Dakolias, Maria. 1995. "A Strategy for Judicial Reform: The Experience in Latin America." *Virginia Journal of International Law* 36.

Davidson, Sinclair, Mikayla Novak, and Jason Potts. 2018. "The $29 Trillion Cost of Trust." *Cryptoeconomics Australia*, July 24.

Davidson, Sinclair, Primavera De Filippi, and Jason Potts. 2018. "Blockchains and the Economic Institutions of Capitalism." *Journal of Institutional Economics* 13 (4).

Day, Matt, Giles Turner, and Natalia Drozdiak. 2019. "Amazon Workers Are Listening to What You Tell Alexa." *Bloomberg*, April 11.

De, Nikhilesh. 2019. "Halt Libra? Us Lawmakers Call for Hearings on Facebook's Crypto." *Coindesk*, June 18.

de Sola Pool, Ithiel. 2009. *Technologies of Freedom*. Cambridge, MA: Harvard University Press.

De Soto, Hernando. 2000. *The Mystery of Capital: Why Capitalism Triumphs in the West and Fails Everywhere Else.* Basic Civitas Books.

Digital Competition Expert Panel. 2019. "Digital Competition Expert Panel: Unlocking Digital Competition." *Government of the United Kingdom*.

Diffie, Whitfield, and Martin Hellman. 1976. "New Directions in Cryptography." *IEEE transactions on Information Theory* 22 (6): 644-654.

Dixit, Avinash K. 2007. *Lawlessness and Economics: Alternative Modes of Governance.* Princeton, New Jersey: Princeton University Press.

Djankov, Simeon et al. 2003. "The New Comparative Economics," *Journal of Comparative Economics* 31 (4).

Domar, Evsey D. 1946. "Capital Expansion, Rate of Growth, and Employment." *Econometrica, Journal of the Econometric Society* 14 (2).

Dorn, James A. 2002. "Economic Development and Freedom: The Legacy of Peter Bauer." *Cato Journal* 22 (2).

Downs, Anthony. 1957. "An Economic Theory of Political Action in a Democracy." *The Journal of Political Economy* 65 (2).

Drexler, Eric K. 2019. "Reframing Superintelligence: Comprehensive AI Services as General Intelligence." In *Technical Report No. 2019-1*. Future of Humanity Institute, University of Oxford.

Du, Juan. 2020. *The Shenzhen Experiment: The Story of China's Instant City*. US: Harvard University Press.

DuBoff, Richard B. 1980. "Business Demand and the Development of the Telegraph in the United States, 1844–1860." *Business History Review* 54 (4): pp. 459-479.

DuPont, Quinn. 2017. "Experiments in Algorithmic Governance: A History and Ethnography of "the Dao," a Failed Decentralized Autonomous Organization." In *Bitcoin and Beyond* edited by Malcolm Campbell-Verduyn. Routledge.

Dunna, Arun, Ciarán O'Brien, and Phillipa Gill. 2018. "Analyzing China's Blocking of Unpublished Tor Bridges." Paper presented at the 8th USENIX Workshop on Free and Open Communications on the Internet (FOCI).

Dwork, Cynthia, and Moni Naor. 1992. "Pricing Via Processing or Combatting Junk Mail." *Paper presented at the Annual International Cryptology Conference, 1992.*

Easterly, William. 2006. "Planners Versus Searchers in Foreign Aid." *Asian Development Review* 23 (2): pp. 1-35.

———. 2006. *The White Man's Burden: Why the West's Efforts to Aid the Rest Have Done So Much Ill and So Little Good*. Penguin Publishing.

Edgar, Timothy H. 2017. *Beyond Snowden: Privacy, Mass Surveillance, and the Struggle to Reform the NSA*. Brookings Institution Press.

Edwards, Lee. 2013. *Leading the Way: The Story of Ed Feulner and the Heritage Foundation*. Crown Publishing Group.

Eichengreen, Barry J. 1995. *Golden Fetters: The Gold Standard and the Great Depression, 1919-1939.* Oxford, UK: Oxford University Press.

Elert, Niklas, and Magnus Henrekson. 2016. "Evasive entrepreneurship." *Small Business Economics* 47 (1): 95-113.

Elert, Niklas, Magnus Henrekson, and Joakim Wernberg. 2016. "Two Sides to the Evasion: The Pirate Bay and the Interdependencies of Evasive Entrepreneurship." *Journal of Entrepreneurship and Public Policy* 5 (2).

Ertel, Wolfgang, and Nathanael T. Black. 2018. *Introduction to Artificial Intelligence.* Springer International Publishing.

Evans, Tonya M. 2019. "Cryptokitties, Cryptography, and Copyright." *American Intellectual Property Law Association Quarterly Journal* 47 (2).

Facebook. 2019. "Facebook Q4 2019 Results." *Facebook.*

Feng, Emily. 2019. "Github Has Become a Haven for China's Censored Internet Users." *NPR*, April 10.

Ferguson, Niall. 2002. *The Cash Nexus: Money and Power in the Modern World, 1700-2000.* New York, New York: Basic Books.

Figes, Orlando. 2008. *The Whisperers: Private Life in Stalin's Russia.* London, UK: Penguin Books.

French, David. 2019. "Before Trump, What Did Conservatism Conserve?" *National Review*, March 8.

Frick, Susanne A., Andrés Rodríguez-Pose, and Michael D. Wong. 2019. "Toward Economically Dynamic Special Economic

Zones in Emerging Countries." *Economic Geography* 95 (1): pp. 30-64.

Friedman, Milton. 1953. "The Methodology of Positive Economics." In *Essays in Positive Economics*, edited by Milton Friedman. Chicago, IL: University of Chicago Press.

Friedman, Patri, and Brad Taylor. 2012. "Seasteading: Competitive Governments on the Ocean." *Kyklos* 65 (2): pp. 217-235.

Fukuyama, Francis. 2006. *The End of History and the Last Man*. New York, NY: Free Press.

Ganne, Emmanuelle. 2018. "Can Blockchain Revolutionize International Trade?" World Trade Organization.

Gans, Joshua. 1998. "Industrialization Policy and the 'Big Push.'" In *Increasing Returns and Economic Analysis*, edited by Kenneth J. Arrow. Palgrave.

Gillespie, Nick, and Matt Welch. 2008. "How 'Dallas' Won the Cold War." *The Washington Post*, April 27.

Goldenfein, Jake, and Andrea Leiter. 2018. "Legal Engineering on the Blockchain: 'Smart Contracts' as Legal Conduct." *Law and Critique* 29 (2).

Good, Irving John. 1966. "Speculations Concerning the First Ultraintelligent Machine." In *Advances in Computers* 6: pp. 31-38.

Gordon, Marcey. "Us Senator Asks Facebook Ceo Mark Zuckerberg to Sell Instagram and Whatsapp." *Business Insider Australia*, September 20.

Governatori, Guido Florian Idelberger, Zoran Milosevic, Regis Riveret, Giovanni Sartor, and Xiwei Xu. 2018. "On Legal Contracts, Imperative and Declarative Smart Contracts, and Blockchain Systems." *Artificial Intelligence and Law* 26 (4): pp. 377–409.

Graglia, J. Michael, and Christopher Mellon. 2018. "Blockchain and Property in 2018: At the End of the Beginning." *Innovations: Technology, Governance, Globalization* 12 (1-2).

Gramlich, Wayne. 1998. "Seasteading – Homesteading the High Seas."

GreatFire.org. 2018.

Green, Matthew. 2019. "Looking Back at the Snowden Revelations." *A Few Thoughts on Cryptographic Engineering*, September 24.

Greenwald, Glenn. 2014. *No Place to Hide: Edward Snowden, the Nsa, and the U.S. Surveillance State.* Henry Holt and Company.

Greif, Avner. 2006. *Institutions and the Path to the Modern Economy: Lessons from Medieval Trade.* UK: Cambridge University Press.

———. 1993. "Contract Enforceability and Economic Institutions in Early Trade: The Maghribi Traders' Coalition." *The American Economic Review* 83 (3).

Gruen, Erich S. 1986. *The Hellenistic World and the Coming of Rome*, vol. 1. Berkeley, CA: University of California Press.

GSMA. 2017. "Blockchain for Development." United Kingdom: GSMA.

Gurri, Adam. 2019. "No One Is Owed an Audience: Facebook, Free Speech, and Liberal Tolerance." *LiberalCurrents*, October 29.

Haas, Benjamin. 2018. "China Bans Winnie the Pooh Film after Comparisons to President Xi." *The Guardian*, August 8.

Halaburda, Hanna, and Miklos Sarvary. 2016. *Beyond Bitcoin: The Economics of Digital Currencies*. Hampshire, UK: Palgrave Macmillan.

Harari, Yuval Noah. 2011. *Sapiens: A Brief History of Humankind.*

Kindle Edition ed.: Random House..

Harnad, Stevan. 2001. "What's Wrong and Right About Searle's Chinese Room Argument?" In *Essays on Searle's Chinese Room Argument* edited by M. Bishop and J. Preston. Oxford University Press.

Harris, Shane. 2013. "The Cowboy of the NSA." *Foreign Policy*, September 9.

Harrod, Roy F. 1939. "An Essay in Dynamic Theory." *The Economic Journal* 49 (193).

Hasan, Haya R., and Khaled Salah. 2019. "Combating Deepfake Videos Using Blockchain and Smart Contracts." *IEEE Access* 7: pp. 41596-41606.

Haugeland, John. 1989. *Artificial Intelligence: The Very Idea*. Cambridge, Massachusetts: MIT Press.

Hausmann, Ricardo, and Dani Rodrik. 2003. "Economic Development as Self-Discovery." *Journal of Development Economics* 72 (2).

Hayek, Friedrich A. 1976. "Choice in Currency: A Way to Stop Inflation." In *The Collective Works of F.A. Hayek: Good Money Part II*, edited by Stephen Kresig. Kent, UK: Tonbridge Printers.

———. 1976. *Denationalisation of Money: The Argument Refined*. London, UKL: Institute of Economic Affairs.

———. 1973. *Law, Legislation and Liberty,* Volume 1: Rules and Order. Chicago, Illinois: University of Chicago Press.

———. 1960 [2011]. *The Fatal Conceit: The Errors of Socialism*. Chicago, Illinois: University of Chicago Press.

———. 1960. *The Constitution of Liberty: The Definitive Edition*. Chicago, Illinois: University of Chicago Press.

———. 1948. *Individualism and Economic Order.* Chicago, Illinois: University of Chicago Press.

———. 1945. "The Use of Knowledge in Society." *The American Economic Review* 35 (4).

———. 1937. "Economics and Knowledge." *Economica* 4 (13).

Hayward, Philip. 2014. "Islands and Micronationality." *Shima* 8 (1).

Heater, Brian. 2019. "The White House Wants to Know If You've Been 'Censored or Silenced' by Social Media." *TechCrunch*, May 16.

Heijden, Jeroen Van der. 2011. "Institutional Layering: A Review of the Use of the Concept." *Politics* 31 (1).

Herman, Edward S. and Noam Chomsky. 2010. *Manufacturing Consent: The Political Economy of the Mass Media*. Penguin Random House.

Hernández, Javier C., and Iris Zhao. 2018. "Students Defiant as Chinese University Warns #Metoo Activist." *The New York Times*, April 24.

Hidalgo, Cesar. 2015. *Why Information Grows: The Evolution of Order, from Atoms to Economies.* Basic Books.

Hilbert, Martin. 2012. "How Much Information Is There in the "Information Society"?" *Significance* 9 (4): pp. 8-12.

———. 2015. "A Review of Large-Scale 'How Much Information' inventories: Variations, Achievements and Challenges." *Information Research* 20 (4).

———. 2017. "Information Quantity." In *Encyclopedia of Big Data* edited by Laurie A. Schintler and Connie L. McNeely. Switzerland: Springer.

Hirschman, Albert O. 1970. *Exit, Voice, and Loyalty: Responses to Decline in Firms, Organizations, and States*, vol. 25. Cambridge, MA: Harvard University Press.

Hiruncharoenvate, Chaya, Zhiyuan Lin, and Eric Gilbert. 2015. "Algorithmically Bypassing Censorship on Sina Weibo with Nondeterministic Homophone Substitutions." *Paper presented at the Ninth International AAAI Conference on Web and Social Media, Oxford, United Kingdom, May 26-29.*

Hobbs, William R., and Margaret E. Roberts. 2018. "How Sudden Censorship Can Increase Access to Information." *American Political Science Review* 112 (3): pp. 621-636.

Holden, Stephen. 2010. "J. R. Ewing Shot Down Communism in Estonia." *The New York Times*, November 11.

Howgego, Christopher J. 1992. "The Supply and Use of Money in the Roman World 200 Bc to Ad 300." *Journal of Roman Studies* 82: pp. 1-31.

———. 1990. "Why Did Ancient States Strike Coins?" *The Numismatic Chronicle* 150: pp. 1-25.

Huang, Ling et al. 2011. "Adversarial Machine Learning." *Paper presented at the Proceedings of the 4th ACM workshop on Security and artificial intelligence, Chicago, Illinois, United States, October 17-21.* See pp. 43–58.

Hukkelås, Håkon, Rudolf Mester, and Frank Lindseth. 2019. "Deepprivacy: A Generative Adversarial Network for Face Anonymization." *arXiv preprint arXiv:1909.04538.*

Hull, Cordell. 1948. *The Memoirs of Cordell Hull*, vol. 1. New York: Macmillan Company.

Humphreys, John. 2016. "Education in Cambodia: Rate of Return and

Personal Equity Finance." *Honours thesis submitted to The University of Queensland.*

James, Henry. 1898. *In the Cage.* London, UK: Duckworth and Co.

Johns, Adrian. 2010. *Death of a Pirate: British Radio and the Making of the Information Age.* WW Norton & Company.

Johnson, Micah K., and Hany Farid. 2005. "Exposing Digital Forgeries by Detecting Inconsistencies in Lighting." *Paper presented at the Proceedings of the 7th Workshop on Multimedia and Security, New York, United States, August 1-2.*

Johnston, Carla B. 2016. *Screened Out: How the Media Control Us and What We Can Do About It: How the Media Control Us and What We Can Do About It.* Routledge.

Johnstone, S. 2011. *A History of Trust in Ancient Greece.* Chicago, Illinois: University of Chicago Press.

Jones, Daniel Stedman. 2012. *Masters of the Universe: Hayek, Friedman, and the Birth of Neoliberal Politics.* Princeton University Press.

Jones, Robert A. 1976. "The Origin and Development of Media of Exchange." *Journal of Political Economy* 84 (4): pp. 757-776.

Kahn, David. 1996. *The Codebreakers: The Comprehensive History of Secret Communication from Ancient Times to the Internet.* New York: Scribner's and Sons.

Kan, Michael. 2013. "GitHub unblocked in China after former Google head slams its censorship." *ComputerWorld*, January 23.

Karpf, David. 2018. "25 Years of Wired Predictions: Why the Future Never Arrives." *Wired*, September 18.

Kelly, Dominic. 2019. *Political Troglodytes and Economic Lunatics:*

The Hard Right in Australia. Melbourne, Australia: Schwartz Publishing.

Kelly, Paul. 1992. *The End of Certainty: The Story of the 1980s.* St Leonards, NSW: Allen & Unwin.

Khanna, Tarun. 1995. "Racing Behavior Technological Evolution in the High-End Computer Industry." *Research Policy* 24 (6): 933-958.

King, Gary, Jennifer Pan, and Margaret E. Roberts. 2013. "How Censorship in China Allows Government Criticism but Silences Collective Expression." *American Political Science Review* 107 (2): pp. 1-18.

Kirzner, Israel M. 1997. "Entrepreneurial Discovery and the Competitive Market Process: An Austrian Approach." *Journal of Economic Literature* 35 (1).

Klein, Naomi. 2007. *The Shock Doctrine: The Rise of Disaster Capitalism.*

Knight, Will. 2017. "The Dark Secret at the Heart of AI." *MIT Technology Review*, April 11.

Kocherlakota, Narayana R. 2002. "Money: What's the Question and Why Should We Care About the Answer?" *American Economic Review* 92 (2): pp. 58-61.

———. 2002. "The Two-Money Theorem." *International Economic Review* 43 (2): pp. 333-346.

———. 1998. "Money Is Memory." *Journal of Economic Theory* 81(2): pp. 232-251.

Komaromi, Ann. 2015. *Uncensored: Samizdat Novels and the Quest for Autonomy in Soviet Dissidence.* Northwestern University Press.

Kuhn, Daniel. "Facebook's New Crypto Faces Scrutiny from European

Authorities." *Coindesk*, June 18.

Kyl, Jon. 2019. "Why Conservatives Don't Trust Facebook." *The Wall Street Journal*, August 20.

LaFleur, Kendal Stephens, and Lei Chen. 2014. "Email Encryption: Discovering Reasons Behind Its Lack of Acceptance." *Paper presented at the Proceedings of the International Conference on Security and Management (SAM), Las Vegas, July 21-24.*

Lapowsky, Issie. 2016. "Here's How Facebook Actually Won Trump the Presidency." *Wired*, November 15.

Leberknight, Chris, and Anna Feldman. 2019. "Leveraging NLP and Social Network Analytic Techniques to Detect Censored Keywords: System Design and Experiments." *Paper presented at the Proceedings of the 52nd Hawaii International Conference on System Sciences, Hawaii, United States, January 8-11.*

Lee, Siu-yau. 2016. "Surviving Online Censorship in China: Three Satirical Tactics and Their Impact." *The China Quarterly* 228: pp. 1061-1080.

Leeson, Peter T. 2009. *The Invisible Hook: The Hidden Economics of Pirates.* USA: Princeton University Press.

———. 2007. "An□Arrgh□Chy: The Law and Economics of Pirate Organization." *Journal of Political Economy* 115 (6): pp. 1049-1094.

Leeson, Peter T., and Peter J. Boettke. 2009. "Two-Tiered Entrepreneurship and Economic Development." *International Review of Law and Economics* 29 (3).

Leng, Sidney, Josh Ye, and Nectar Gan. 2017. "The Who, What and Why in China's Latest VPN Crackdown." *South China Morning Post*, January 26.

Leoni, Bruno. *Freedom and the Law*. 1961. The William Volker Fund Series in the Humane Studies. Princeton, New Jersey: Van Nostrand.

Lepp, Annika, and Mervi Pantti. 2013. "Window to the West: Memories of Watching Finnish Television in Estonia During the Soviet Period." *View Journal of European Television History and Culture* 1 (3): 77-87.

Lewis, W. Arthur. 1954. "Economic Development with Unlimited Supplies of Labour." *The Manchester School* 22 (2).

Libra white paper is available at https://libra.org/en-US/.

Locard, H. 2004. *Pol Pot's Little Red Book: The Sayings of Angkar.* Chiang Mai, Thailand: Silkworm Books.

Lustig, Eileen. 2009. "Money Doesn't Make the World Go Round: Angkor's Non-Monetisation." In *Research in Economic Anthropology* 29: *"Economic Development, Integration, and Morality in Asia and the Americas."* ed. Donald C. Wood. Bingley, UK: Emerald Group Publishing.

Lutter, Mark. 2017. "The Case for Innovative Governance." *Centre for Innovative Governance Research.*

———. 2016. "Three Essays on Proprietary Cities." *George Mason University.*

MacDonald, Trent J. 2019. *The Political Economy of Non-Territorial Exit.* Edward Elgar Publishing.

———. 2015. "Theory of Non-Territorial Internal Exit." *SSRN 2661226.*

MacDonald, Trent J., Darcy W.E. Allen, and Jason Potts. 2016. "Blockchains and the Boundaries of Self-Organized Economies: Predictions for the Future of Banking." In *Banking*

Beyond Banks and Money: A Guide to Banking Services in the Twenty-First Century, edited by Paolo Tasca, et al. Springer International Publishing.

Madani, Dorsati. 1999. "A Review of the Role and Impact of Export Processing Zones." Washington, DC: The World Bank.

Mance, Henry. 2019. "Is Privacy Dead?" *Financial Times*, July 19.

Manezhev, Sergei. 1993. "Free Economic Zones in the Context of Economic Changes in Russia." *Europe-Asia Studies* 45 (4).

Manners, Ron. "Liberty could be good for you too!" *Speech at Mont Pelerin Society, Gran Canaria, Spain, October 23, 2018.*

Maras, Marie-Helen, and Alex Alexandrou. 2019. "Determining Authenticity of Video Evidence in the Age of Artificial Intelligence and in the Wake of Deepfake Videos." *The International Journal of Evidence & Proof* 23 (3): pp. 255-262.

Marglin, Stephen A. 1974. "What Do Bosses Do? The Origins and Functions of Hierarchy in Capitalist Production." *Review of Radical Political Economics* 6 (2): pp. 60-112.

Marx, Karl. 1867. *Capital: A Critique of Political Economy*. vol. 1. Germany.

Mauro, Paolo. 1995. "Corruption and Growth." *The Quarterly Journal of Economics* 110 (3).

May, Timothy C. 1992. "The Crypto Anarchist Manifesto." November 22.

McChesney, Robert Waterman, Robert W. McChesney, and John Nichols. 2002. *Our Media, Not Theirs: The Democratic Struggle against Corporate Media.* New York: Seven Stories Press.

McKinney, Scott A., Rachel Landy, and Rachel Wilka. 2017. "Smart Contracts, Blockchain, and the Next Frontier of Transactional

Law." *Washington Journal of Law, Technology Arts* 13.

McKinnon, Ronald I. 1996. *The Rules of the Game: International Money and Exchange Rates.* Cambridge, Massachusetts: MIT Press.

McLaughlin, Jenna. 2016. "Spy Chief Complains That Edward Snowden Sped up Spread of Encryption by 7 Years." *The Intercept*, April 26.

McMurren, Juliet, Andrew Young, and Stefaan Verhulst. 2019. "Addressing Transaction Costs through Blockchain and Identity in Swedish Land Transfers." *GovLab.*

McPherson, Richard, Reza Shokri, and Vitaly Shmatikov. 2016. "Defeating Image Obfuscation with Deep Learning." *arXiv preprint arXiv:1609.00408.*

Mehar, Muhammad Izhar, Charles Louis Shier, Alana Giambattista, Elgar Gong, Gabrielle Fletcher, Ryan Sanayhie, Henry M Kim, and Marek Laskowski. 2019. "Understanding a Revolutionary and Flawed Grand Experiment in Blockchain: The Dao Attack." *Journal of Cases on Information Technology* (JCIT) 21 (1): pp. 19-32.

Menefee, Samuel Pyeatt. 1994. "Republics of the Reefs: Nation-Building on the Continental Shelf and in the World's Oceans." *California Western International Law Journal* 25 (1).

Merkle, Ralph C. 1988. "A Digital Signature Based on a Conventional Encryption Function." In *Advances in Cryptology — Crypto '87:* pp. 369-378. Berlin, Germany: Springer.

Middleton, John. 2018. "The Rise of the International Commercial Court." In *The 2018 Hong Kong International Commercial Law Conference, Hong Kong.*

Milgrom, Paul R., Douglass C. North, and Barry R. Weingast. 1990. "The Role of Institutions in the Revival of Trade: The Law Merchant, Private Judges, and the Champagne Fairs." *Economics & Politics* 2 (1).

Mill, John Stuart. 1859. *On Liberty*. London, UK: J. W. Parker and Son

Miller, Andrew. 2016. "Ethereum Isn't Turing Complete and It Doesn't Matter Anyway." *Youtube*, August 7.

Miller, Chris. 2016. *The Struggle to Save the Soviet Economy: Mikhail Gorbachev and the Collapse of the USSR.* US: University of North Carolina Press Books.

Miller, Mark S., and Marc Stiegler. 2003. "The Digital Path: Smart Contracts and the Third World." In *Markets, Information and Communication*, edited by Jack Birner and Pierre Garrouste. Routledge.

Mirowski, Philip, and Dieter Plehwe. 2015. *The Road from Mont Pèlerin.* Cambridge, MA: Harvard University Press.

Mises, Ludwig von. 1949. *Human Action: A Treatise on Economics.* New York: Fox & Wilkes.

———. 1953. *The Theory of Money and Credit.* Auburn, Alabama: Yale University Press.

———. 1998. *Human Action: A Treatise on Economics*. Auburn, Alabama: Yale University Press.

Mnookin, Robert H., and Lewis Kornhauser. 1979. "Bargaining in the Shadow of the Law: The Case of Divorce." *The Yale Law Journal* 88 (5): pp. 950-997.

Moberg, Lotta. 2015. "The Political Economy of Special Economic Zones." *Journal of Institutional Economics* 11 (1): pp. 167-190.

Moberg, Lotta, and Vlad Tarko. 2014. "Why No Chinese Miracle in Africa? Special Economic Zones and Liberalization Avalanches." In *Special Economic Zones and Liberalization Avalanches*.

Monti, Federico, et al. 2019. "Fake News Detection on Social Media Using Geometric Deep Learning." *arXiv preprint arXiv:1902.06673*.

Montinola, Gabriella, Yingyi Qian, and Barry R Weingast. 1995. "Federalism, Chinese Style: The Political Basis for Economic Success in China." *World Politics* 48 (1).

Morgan, Jacob. 2014. "Privacy Is Completely and Utterly Dead, and We Killed It." *Forbes*, August 19.

Morozov, Evgeny. 2013. *To Save Everything, Click Here: The Folly of Technological Solutionism.* PublicAffairs eBook.

Moyer, Thomas J., and Emily Stewart Haynes. 2002. "Mediation as a Catalyst for Judicial Reform in Latin America." *Ohio State Journal on Dispute Resolution* 18 (3).

Muhammad Izhar Mehar et al. 2019. "Understanding a Revolutionary and Flawed Grand Experiment in Blockchain: The Dao Attack." *Journal of Cases on Information Technology (JCIT)* 21 (1).

Munger, Michael C. 2018. *Tomorrow 3.0: Transaction Costs and the Sharing Economy*. Cambridge, UK: Cambridge University Press.

Murphy, Katherine. 2011. "We Need to Lift Our Game, Press Watchdog Says." *The Sydney Morning Herald*, November 9.

Murphy, Kevin M., Andrei Shleifer, and Robert W. Vishny. 1989. "Industrialization and the Big Push." *The Journal of Political Economy* 97 (5).

Nakamoto, Satoshi. 2008. "Bitcoin: A Peer-to-Peer Electronic Cash

System." *Bitcoin.org.*

Nasr, Milad, Sadegh Farhang, Amir Houmansadr and Jens Grossklags. 2019. "Enemy at the Gateways: Censorship-Resilient Proxy Distribution Using Game Theory." *Paper presented at the Network and Distributed Systems Security (NDSS) Symposium, San Diego, California, United States, 24-27 February.*

Neff, Robert. 2007. "Flags That Hide the Dirty Truth." *Asia Times*, April 19.

Nelson, Richard R. 1959. "The Simple Economics of Basic Scientific Research." *Journal of Political Economy* 67 (3).

Neumeyer, Ken. 1982. *Sailing the Farm.* Ten Speed Press.

Neustaedter, Carman, Saul Greenberg, and Michael Boyle. 2006. "Blur Filtration Fails to Preserve Privacy for Home-Based Video Conferencing." *ACM Transactions on Computer-Human Interaction (TOCHI)* 13 (1).

Newman, Lily Hay. 2018. "Australia's Encryption-Busting Law Could Impact Global Privacy." December 7.

Newzoo. 2019. "Global Mobile Market Report." *Newzoo.*

Ng, Kei Yin, Anna Feldman, Jing Peng and Chris Leberknight. 2019. "Neural Network Prediction of Censorable Language." *Paper presented at the Proceedings of the Third Workshop on Natural Language Processing and Computational Social Science, Minneapolis, Minnesota, June 6.*

Ng, Kei Yin, Anna Feldman, and Chris Leberknight. 2018. "Detecting Censorable Content on Sina Weibo: A Pilot Study." *Paper presented at the Proceedings of the 10th Hellenic Conference on Artificial Intelligence, Patras, Greece, July 9-12.*

Noah Harari, Yuval. 2011. *Sapiens: A Brief History of Humankind.*

Kindle Edition ed., Random House.

North, Douglass C., John Joseph Wallis, and Barry R Weingast. 2009. *Violence and Social Orders: A Conceptual Framework for Interpreting Recorded Human History.* UK: Cambridge University Press.

Novak, Mikayla. 2018. "Crypto-Altruism: Some Institutional Economic Considerations." *SSRN 3230541*.

Novak, Mikayla. (forthcoming) "Crypto-Friendliness: Understanding Blockchain Public Policy." *Journal of Entrepreneurship and Public Policy*.

Novak, Mikayla, Sinclair Davidson, and Jason Potts. 2018. "The Cost of Trust: A Pilot Study." *Journal of the British Blockchain Association* 1 (2): pp. 1-7.

OECD. 2020. "Social Expenditure - Aggregated Data." *OECD.Stat*.

OECD. n.d. "Development Aid Stable in 2017 with More Sent to Poorest Countries."

O'Flaherty, Kate. 2019. "Who Is the Dark Overlord Threatening to Leak Sensitive 9/11 Documents?" *Forbes*, January 2.

O'Shields, Reggie. 2017. "Smart Contracts: Legal Agreements for the Blockchain." *North Carolina Banking Institute* 21.

Olson, Mancur. 2009. *The Logic of Collective Action*, vol. 124. Cambridge, Massachusetts: Harvard University Press.

Ostrom, Elinor. 1990. *Governing the Commons: The Evolution of Institutions for Collective Action.* Cambridge, UK: Cambridge University Press.

Page, Larry, and Sergey Brin. 2004. "Letter from the Founders: 'An Owner's Manual' for Google Shareholders." August 18.

Parkin, Siodhbhra, and Jiayun Feng. 2019. "China's #Metoo Movement, Explained." *SupChina*, July 12.

Pearl, Judea. 2019. "The Limitations of Opaque Learning Machines." In *Possible Minds: Twenty-Five Ways of Looking at Ai*, edited by John Brockman.

Peltzman, Sam. 1976. "Toward a More General Theory of Regulation." In *National Bureau of Economic Research Working Paper 133.*

Petersen, Julie K. 2007. *Understanding Surveillance Technologies: Spy Devices, Their Origins & Applications.* New York: Auerbach Publications.

Peterson, Christine, Mark S. Miller, and Allison Duettmann. 2017. "Cyber, Nano, and AGI Risks: Decentralized Approaches to Reducing Risks." *Paper presented at the Colloquium on Catastrophic and Existential Risk, University of California, Los Angeles, United States, March 27-29.*

Peterson, Jordan. 2019. "The Deepfake Artists Must Be Stopped before We No Longer Know What's Real." *National Post*, August 23.

Pisa, Michael, and Matt Juden. 2017. "Blockchain and Economic Development: Hype Vs. Reality." *Center for Global Development Policy,* Paper No. 107.

Poblet, Marta. 2017. "Inside Catalonia's Cypherpunk Referendum." *Eureka Street* 27 (20), October 6.

Posner, Eric A., and E. Glen Weyl. 2018. *Radical Markets: Uprooting Capitalism and Democracy for a Just Society*. New Jersey, US: Princeton University Press.

Potts, Jason, and Ellie Rennie. 2019. "Web3 and the Creative Industries:

How Blockchains Are Reshaping Business Models." Chapter 6 in *A Research Agenda for Creative Industries*, edited by S. Cunningham. Cheltenham, UK: Edward Elgar Publishing.

Potts, Jason, Ellie Rennie, and Jake Goldenfein. 2017. "Blockchains and the Crypto City." *it-Information Technology* 59 (6): pp. 285-293.

Potts, Jason, John Humphreys, and Joseph Clark. 2018. "A Blockchain-Based Universal Basic Income." *Medium*, January 16.

Qiang, Xiao. 2004. "A Bold New Voice - Lu Yuegang's Extraordinary Open Letter to Authorities." *China Digital Times*, July 20.

Quinn, Jameson. 2013. "Matching $ Experiment: 10x Your Impact, or I'll Give You $5 Free." *Quora*, August 18.

Quirk, Joe, and Patri Friedman. 2017. *Seasteading: How Floating Nations Will Restore the Environment, Enrich the Poor, Cure the Sick, and Liberate Humanity from Politicians.* Simon and Schuster.

Radio New Zealand. 2018. "Hundreds March in Tahiti Against Building of Floating Islands." *Radio New Zealand*, April 9.

Rana, Gaurav. 2018. "Timechain: A Decade of Misunderstanding Blockchain." *Good Audience*, October 9.

Rand, Ayn. 2005. *Atlas Shrugged.* Penguin Books.

Raskin, Max. 2017. "The Law and Legality of Smart Contracts." *Georgetown Law Technology Review* 1 (2): pp. 305-341.

Ratcliffe, Jonathan. 2019. "How Many CCTV Cameras Are There in London 2019?" *CCTV.co.uk*, May 29.

Read, Max. 2016. “Donald Trump Won Because of Facebook.” *New York*, November 9.

Reger, Gary. 1994. *Regionalism and Change in the Economy of Independent Delos*. Berkeley, CA: University of California Press Berkeley.

Rennie, Ellie, Jason Potts, and Ana Pochesneva. 2019. “Blockchain and the Creative Industries.” In *RMIT Blockchain Innovation Hub Provocation Paper*.

Richburg, Keith B. 2011. “Nervous About Unrest, Chinese Authorities Block Web Site, Search Terms.” *The Washington Post*, February 25.

Roberts, Margaret E. 2018. *Censored: Distraction and Diversion inside China's Great Firewall*. New Jersey, United States: Princeton University Press.

Roberts, Hal, Ethan Zuckerman, Jillian York, Robert Faris, and John Palfrey. 2010. “2010 Circumvention Tool Usage Report.” *The Berkman Center for Internet & Society*, October.

Rodrik, Dani. 2008. “The New Development Economics: We Shall Experiment, but How Shall We Learn?” *Harvard Kennedy School Working Paper Series*.

Rosenstein-Rodan, Paul N. 1943. “Problems of Industrialisation of Eastern and South-Eastern Europe.” *The Economic Journal* 53 (210/211).

Rössler, Andreas, et al. 2018. “Faceforensics: A Large-Scale Video Dataset for Forgery Detection in Human Faces.” *arXiv preprint arXiv:1803.09179*.

Rossetto, Louis. 1993. “The Wired Manifesto.” *Wired*.

Rostow, Walt Whitman. 1990. *The Stages of Economic Growth: A Non-Communist Manifesto*. Cambridge University Press.

Rothman, Noah. 2016. "What Conservatism Has Conserved." *Commentary*, October 18.

Rule, Sheila. 1989. "Reagan Gets a Red Carpet from the British." *The New York Times*, June 14.

Ryan, Philippa. 2017. "Smart Contract Relations in E-Commerce: Legal Implications of Exchanges Conducted on the Blockchain." *Technology Innovation Management Review* 7 (10): pp. 14-21.

Sachs, Jeffrey. 2005. *The End of Poverty: How We Can Make It Happen in Our Lifetime* UK: Penguin Publishing.

Safaka, Iris, Christina Fragouli, and Katerina Argyraki. 2016. "Matryoshka: Hiding Secret Communication in Plain Sight." *Paper presented at the 6th USENIX Workshop on Free and Open Communications on the Internet (FOCI), Austin, Texas, August 6.*

Sanchez, Julian. 2014. "Snowden: Year One." *Cato Unbound*, June 5.

Schaps, David. 2004. *The Invention of Coinage and the Monetization of Ancient Greece*. Ann Arbor, Michigan: University of Michigan Press.

Schatz, Brian, and Sherrod Brown. 2019. "Letter to Mr Patrick Collison." *Published letters, October 8.*

Schmidtz, David. 2006. "I'm Not a Utilitarian, but I Play One on TV." *Cato Unbound*, March 20.

Schrepel, Thibault. 2020. "The Theory of Granularity: A Path for Antitrust in Blockchain Ecosystems." *SSRN* Paper No. 3519032.

Searle, John R. 1980. "Minds, Brains, and Programs." *Behavioral and Brain Sciences* 3 (3): 417-457.

Selgin, George, and Lawrence H White. 1999. "A Fiscal Theory of Government's Role in Money." *Economic Inquiry* 37 (1): pp.

154-165.

Shahbaz, Adrian. 2018. "Freedom on the Net 2018: The Rise of Digital Authoritarianism." *Freedom House*.

Shankland, Stephen. 2008. "Google Begins Blurring Faces in Street View." *CNET*, May 13.

Shen, Fei, and Zhi'an Zhang. 2018. "Do Circumvention Tools Promote Democratic Values? Exploring the Correlates of Anti-Censorship Technology Adoption in China." *Journal of Information Technology & Politics* 15 (2).

Shermin, Voshmgir. 2017. "Disrupting Governance with Blockchains and Smart Contracts." *Strategic Change* 26 (5): pp. 499-509.

Shleifer, Andrei. 2005. "Understanding Regulation." *European Financial Management* 11 (4).

Shleifer, Andrei, and Robert W. Vishny. 1993. "Corruption." ibid.108

Shu, Kai, Amy Sliva, Suhang Wang, Jiliang Tang, and Huan Liu. 2017. "Fake News Detection on Social Media: A Data Mining Perspective." *ACM SIGKDD Explorations Newsletter* 19 (1).

Sills, Kate. 2018. "Code Is Capital." *libertarianism.org*, May 9.

Skilling, Harold Gordon. 1989. *Samizdat and an Independent Society in Central and Eastern Europe*. Macmillan Press, in association with St Antony's College, Oxford.

Snow, Marcellus S. 2002. "Competition as a Discovery Procedure." *Quarterly Journal of Austrian Economics* 5 (3): pp. 9–23.

Snowden, Edward. 2019. *Permanent Record.* UK: Pan Macmillan.

Somin, Ilya. 2016. *Democracy and Political Ignorance: Why Smaller Government Is Smarter*. Stanford University Press.

———. 2013. "Democracy and Political Ignorance." *Cato Unbound*, October 11.

Song, Jimmy. 2018. "The Truth About Smart Contracts." *Medium*, June 11.

Standage, Tom. 1998. *The Victorian Internet.* New York, US: Bloomsbury Publishing

Stigler, George J. 1971. "The Theory of Economic Regulation." *The Bell Journal of Economics and Management Science* 2 (1): pp. 3-21.

Strickland, Eliza. 2018. "AI-Human Partnerships Tackle 'Fake News'." *IEEE Spectrum*, August 29.

Stringham, Edward Peter. 2011. *Anarchy and the Law: The Political Economy of Choice*, vol. 1. Transaction Publishers.

Strong, Michael, and Robert Himber. 2009. "The Legal Autonomy of the Dubai International Financial Centre: A Scalable Strategy for Global Free-Market Reforms." *Economic Affairs* 29 (2): pp. 36-41.

Suzor, Nicolas P., Kylie M. Pappalardo, and Natalie McIntosh. 2017. "The Passage of Australia's Data Retention Regime: National Security, Human Rights, and Media Scrutiny." *Internet Policy Review* 6 (1).

Swan, Trevor W. 1956. "Economic Growth and Capital Accumulation." *Economic Record* 32 (2).

Swire, Peter. 2015. "Going Dark: Encryption, Technology, and the Balance between Public Safety and Privacy." *Hearing by Senate Judiciary Committee*, July 8.

Szabo, Nick. 1997. "The Idea of Smart Contracts." *Nick Szabo's Papers and Concise Tutorials*.

Tabarrok, Alex. 2013. "A Test of Dominant Assurance Contracts." *Marginal Revolution*, August 29.

———. 1998. "The Private Provision of Public Goods Via Dominant Assurance Contracts." *Public Choice* 96 (3-4): pp. 345-362.

Tallinn, Jaan. 2017. "AI and Value Alignment." In *Beneficial AI.*

Tanzi, Vito, and Ludger Schuknecht. 2000. *Public Spending in the 20th Century: A Global Perspective.* Cambridge, Massachusetts: Cambridge University Press.

Teles, Steven. 2013. "Kludgeocracy in America." *National Affairs* 17 (Fall).

Tene, Omer. 2008. "What Google Knows: Privacy and Internet Search Engines." *Utah Law Review*: pp. 1433-54.

Thakur, Vinay et al. 2019. "Land Records on Blockchain for Implementation of Land Titling in India." *International Journal of Information Management.*

TheDarkOverlord. "Press Release 02 – Crypto-Cash for Crypto-Cache."

The Gentleman's Magazine. 1857. "A New Calculating Machine." *The Gentleman's Magazine.*

The Global Times. 2015. "What Impact Does the Firewall Bring to the Chinese Internet?" *The Global Times*, January 28.

The Washington Post. 2019. "Deepfakes Are Dangerous — and They Target a Huge Weakness." *The Washington Post*, June 17.

The White House. "White House Type Form."

Thierer, Adam. 2018. "How Well-Intentioned Privacy Regulation Could Boost Market Power of Facebook & Google." *The Technology Liberation Front*, April 25.

———. 2018. "GDPR Compliance: The Price of Privacy Protections." *The Technology Liberation Front*, July 9.

———. 2014. *Permissionless Innovation: The Continuing Case for Comprehensive Technological Freedom.* Mercatus Center at George Mason University.

———. 2005. *Media Myths: Making Sense of the Debate over Media Ownership*. Washington, DC: The Progress & Freedom Foundation.

Tiebout, Charles M. 1956. "A Pure Theory of Local Expenditures." *Journal of Political Economy* 64 (5): pp. 416-424.

Trakman, Leon E. 1983. *The Law Merchant: The Evolution of Commercial Law.* Littleton, Colorado: William S. Hein & Co.

Trumbull, Robert. 1972. "Pacific Islanders Fight Reef Plan." *The New York Times*, February 27.

Tucker, Jeffrey. 2015. *Bit by Bit: How P2P is Freeing the World.* US: Liberty.me.

Tullock, Gordon. 1975. "The Transitional Gains Trap." *The Bell Journal of Economics* 6 (2).

Turing, Alan M. 1950. "Computing Machinery and Intelligence." *Mind* 59 (236): 433-460.

Turner, Jacob. 2019. *Robot Rules: Regulating Artificial Intelligence.* Palgrave Macmillan.

Twitter. 2019. "Twitter Selected Company Metrics and Financials." *Twitter*.

United Nations. 2005. "Report of the Fourth Meeting of the Judicial Integrity Group." Vienna, Austria: United Nations Office on Drugs and Crime.

Urbina, Ian. 2019. *The Outlaw Ocean: Journeys across the Last Untamed Frontier.* Knopf Doubleday Publishing Group.

US House of Representatives Intelligence Committee. 2019. "House Intelligence Committee to Hold Open Hearing on Deepfakes and AI: The National Security Challenge of Artificial Intelligence, Manipulated Media, and 'Deepfakes.'" *House Intelligence Committee* news release, 7 June.

Vanberg, Viktor, and Wolfgang Kerber. 1994. "Institutional Competition among Jurisdictions: An Evolutionary Approach." *Constitutional Political Economy* 5 (2): pp. 193-219.

Vikhoreva, Svetlana J. 2001. "The Development of Free Economic Zones in Russia." *The Economic Research Institute for Northeast Asia (ERINA) Report* 38.

Vitalik.eth. 2018. "To be clear, at this point I quite regret adopting the term "smart contracts". I should have called them something more boring and technical, perhaps something like "persistent scripts"." *Twitter*, October 13, 2018.

Volokh, Eugene. 1996. "Freedom of Speech in Cyberspace from the Listener's Perspective: Private Speech Restrictions, Libel, State Action, Harassment, and Sex." *The University of Chicago Legal Forum.*

Wagner, Richard E. 2016. "The Peculiar Business of Politics." *In GMU Working Paper in Economics No. 16-27.*

———. 2016. *Politics as a Peculiar Business: Insights from a Theory of Entangled Political Economy.* Cheltenham, UK: Edward Elgar Publishing.

Walker, Jesse. 2004. *Rebels on the Air: An Alternative History of Radio in America.* New York, US: New York University Press.

Wallace, Kurt. 2019. *Regulatory Dark Matter: How Unaccountable Regulators Subvert Democracy by Imposing Red Tape without Transparency.* Institute of Public Affairs, June.

Wang, Jin. 2013. "The Economic Impact of Special Economic Zones: Evidence from Chinese Municipalities." *Journal of Development Economics* 101.

Warren, Elizabeth. "Here's How We Can Break up Big Tech." *Medium,* March 8.

Weingast, Barry R. 1995. "The Economic Role of Political Institutions: Market-Preserving Federalism and Economic Development." *Journal of Law, Economics & Organization* 11 (1).

Weiwei, Ai. 2013. *Weiwei-Isms.* Princeton, New Jersey: Princeton University Press.

Werbach, Kevin. 2018. *The Blockchain and the New Architecture of Trust.* Cambridge, MA: MIT Press.

Werbach, Kevin, and Nicolas Cornell. 2017. "Contracts Ex Machina." *Duke Law Journal* 67: pp. 313-382.

Westin, Alan F. 1967. *Privacy and Freedom.* New York: Atheneum.

Whitaker, Amy, and Roman Kräussl. 2018"Blockchain, Fractional Ownership, and the Future of Creative Work." *CFS Working Paper Series.*

White, Lawrence H. 2007. "Payments System Innovations in the United States since 1945 and Their Implications for Monetary Policy." In *Institutional Change in the Payments System and Monetary Policy*, edited by Stefan W. Schmitz and Geoffrey Wood. Routledge.

Williamson, Oliver E. 1988. "Corporate Finance and Corporate Governance." *The Journal of Finance* 43 (3): pp. 567-591.

———. 1985. *The Economic Institutions of Capitalism*. New York: Free Press.

Winter, Philipp, and Stefan Lindskog. 2012. *How the Great Firewall of China Is Blocking Tor.* USENIX-The Advanced Computing Systems Association.

Woodard, George W. 2010. "Cold War Radio Jamming." In *Cold War Broadcasting: Impact on the Soviet Union and Eastern Europe, a Collection of Studies and Documents*, edited by Ross Johnson and Gene Parta. Budapest and New York: CEU Press.

Wright, Aaron, and Primavera De Filippi. 2018. *Blockchain and the Law: The Rule of Code*. Cambridge, Massachusetts: Harvard University Press.

Wurman, Richard Saul. 1989. *Information Anxiety.* California, United State: Doubleday.

Yang, Yuan. 2018. "Artificial Intelligence Takes Jobs from Chinese Web Censors." *The Financial Times*, May 22.

Yeung, Yue-man, Joanna Lee, and Gordon Kee. 2009. "China's Special Economic Zones at 30." *Eurasian Geography and Economics* 50 (2): pp. 222– 240.

York, Jillian. 2018. "Facebook: The New Censor's Office." *New Internationalist*, April 17.

Yudkowsky, Eliezer. 2016. "The AI Alignment Problem: Why It's Hard, and Where to Start." In *Symbolic Systems Distinguished Speaker series,* Stanford University, May 5.

Zakharov, Egor, Aliaksandra Shysheya, Egor Burkov, and Victor Lempitsky. 2019. "Few-Shot Adversarial Learning of Realistic Neural Talking Head Models." *arXiv preprint arXiv:1905.08233.*

Zegart, Amy B. 2006. "An Empirical Analysis of Failed Intelligence

Reforms before September 11." *Political Science Quarterly* 121 (1): pp. 33-60.

Zhang, Han. 2018. "One Year of #Metoo: How the Movement Eludes Government Surveillance in China." *The New Yorker*, October 10.

Zhao, Wolfie. 2018. "#Metoo Movement Turns to Ethereum to Evade Censorship." *CoinDesk*, April 25.

Zhou, Xinyi, and Reza Zafarani. 2018. "Fake News: A Survey of Research, Detection Methods, and Opportunities." *arXiv preprint arXiv:1812.00315*.

Zuckerberg, Mark. 2019. "Testimony to Congress." *Hearing Before the United States House of Representatives Committee on Financial Services*.

ABOUT THE AUTHORS

Dr Darcy W.E. Allen is a Research Fellow at the RMIT Blockchain Innovation Hub. He is an economist and writer working on the economics of new technologies. Dr Allen has published across law and economics including in journals such as Research Policy and the Harvard Negotiation Review. His recent books are *Unfreeze: How to Create a High Growth Economy After the Pandemic* (AIER, 2020), and *Cryptodemocracy: How Blockchain Can Radically Expand Democratic Choice* (Lexington, 2019). He has appeared many times as an expert witness before Australian parliamentary inquiries, his opinion pieces and commentary have appeared widely across the electronic and print media, and he sits on the editorial board of several blockchain journals. Twitter: @DrDarcyAllen

Dr Chris Berg is a Senior Research Fellow and Co-Director of the RMIT Blockchain Innovation Hub, the world's first dedicated social science research centre studying blockchain technology, based at RMIT University, Melbourne. Dr Berg is author of ten books, including *The Libertarian Alternative* (Melbourne University Press), *Understanding the Blockchain Economy: An Introduction to Institutional Cryptoeconomics* (Edward Elgar) and The Classical Liberal Case for Privacy in a World of Surveillance and Technological Change (Palgrave Macmillan). He is an Adjunct Fellow with the Institute of Public Affairs, an Academic Fellow with the Australian Taxpayers' Alliance, and a Research Fellow with the University College London Centre for Blockchain Technologies.

Professor Sinclair Davidson is Professor of Institutional Economics at the RMIT Blockchain Innovation Hub at RMIT University, an Adjunct Fellow at the Institute of Public Affairs, an Academic Fellow at the Australian Taxpayers' Alliance, an Adjunct Economics Fellow at the Consumer Choice Center, and a Research Associate at the University College London Centre for Blockchain Technologies. He is a member of the Centre for Independent Studies Council of Academic Advisers. Sinclair has published in academic journals such as the *European Journal of Political Economy*, *Journal of Economic Behavior and Organization*, *Economic Affairs*, and *The Cato Journal*. He is a regular contributor to public debate. His opinion pieces have been published in *The Age*, *The Australian*, *Australian Financial Review*, *The Conversation*, *Daily Telegraph*, *Sydney Morning Herald*, and *Wall Street Journal Asia*. He blogs at Catallaxy Files and Tweets @SincDavidson and @Cryptoeconomico.

ABOUT AIER

The American Institute for Economic Research in Great Barrington, Massachusetts, was founded in 1933 as the first independent voice for sound economics in the United States. Today it publishes ongoing research, hosts educational programs, publishes books, sponsors interns and scholars, and is home to the world-renowned Bastiat Society and the highly respected Sound Money Project. The American Institute for Economic Research is a 501c3 public charity.

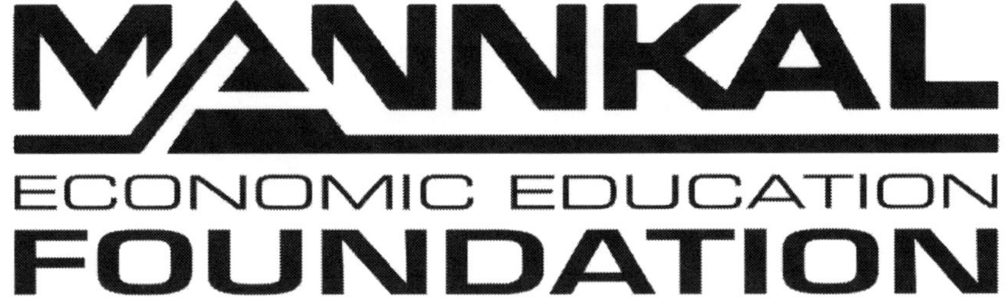

The Mannkal Economic Foundation is a free market think tank based in Perth, Western Australia.

We believe that the principles of limited government and free markets are essential components for the future success of Australia.

Individals around the world have benefitted from the unparalleled success of capitalism, which continues to be under attack. It is up to us to defend it in the marketplace of ideas.

Our flagship activity is our Leadership Development Program (LDP), which provides scholarships to Western Australianh university students to attend seminars, attend conferences and travel internationally, experiencing the freedom movement around the world.

Mannkal's research focuses on Western Australian issues. We apply insights from classical liberal thinkers to produce sound policy recommendations at the state level. Our program scholars are actively incolced in this work.

In delivering the LDP, we 'awaken opporutnities' in the next generation, providing the basis for human flourishing.

For further information, please contact:

Mannkal Economic Education Foundation
'Hayek on Hood'
3/31 Hood Street
Subiaco, Western Australia 6008
AUSTRALIA

PHONE: +61 9382 1288
EMAIL: mannwest@mannkal.org
WEBSITE: www.mannkal.org

To be added to our newsletter please visit our website.

Made in the USA
Middletown, DE
19 August 2020